Isley
Eisley

Oui
Oui

Dear sista!
Look out for the boys... they are grosses. They will stab you AS you walk away... AND THEN they will be OBNOXIOUS!

So if you see a boy RUN AWAY! FAST!

MATHEMATICAL MODELING

BOOK 1

AUTHORS

Maurice Barry
Marian Small
Anne Avard-Spinney
Linda Lesley Burnard Wheadon

Provincial Mathematics Consultants

New Brunswick
John Hildebrand
Greta Gilmore

Newfoundland and Labrador
Patricia Maxwell
Sadie May

Nova Scotia
Richard MacKinnon

Review Panel

David DeCoste
St. Francis Xavier University, Nova Scotia

Roddie J.A. Duguay
Rothesay High School, New Brunswick

Dennis Ivany
Botwood Collegiate, Newfoundland

Hugh McKnight
St. Stephen High School, New Brunswick

Ralph Montesanto
Saltfleet District High School, Ontario

Lynn Phillips
Park View Education Centre, Nova Scotia

Edward J. Somerton
Provincial Mathematics Consultant (retired)
Newfoundland

Leanne Zorn
Mount Boucherie Secondary School,
British Columbia

Australia • Canada • Denmark • Japan • Mexico • New Zealand • Philippines
Puerto Rico • Singapore • South Africa • Spain • United Kingdom • United States

1120 Birchmount Road
Scarborough, Ontario M1K 5G4
www.nelson.com
www.thomson.com

Copyright © 2000 Nelson, a division of Thomson Learning.
Thomson Learning is a trademark used herein under license.

ALL RIGHTS RESERVED. No part of this work covered by the copyright hereon may be reproduced, transcribed, or used in any form or by any means—graphic, electronic, or mechanical, including photocopying, recording, taping, Web distribution or information storage and retrieval systems—without the permission of the publisher.

For permission to use material from this text or product, contact us by
- Web: www.thomsonrights.com
- Phone: 1-800-730-2214
- Fax: 1-800-730-2215

Canadian Cataloguing in Publication Data

Barry, Maurice
    Mathematical modeling, book 1

For use in Atlantic Canada.
Includes index.
ISBN 0-176-05981-4

1. Mathematics. I. Spinney, Anne, 1949–
II. Small, Marian. III. Title.

QA39.2.M37 1999   510   C99-931779-2

Publisher, Mathematics	Cheryl Turner
Mathematics Consultant	David Zimmer
Project Manager	Colin Garnham
Senior Project Editor	Winnie Siu
Editors	Diane Brassolotto, Mary Reeve, Laurel Sparrow, Jackie Williams
Art Director	Angela Cluer
Series Design	Sue Peden
Cover Design	Monica Kompter
Composition Analyst	Daryn DeWalt, Nelson Gonzales
Photos and Permissions	Vicki Gould
Technical Art	Patricia Code, Deborah Crowle, Irma Ikonen, Tom De Ritter
APEF Regional Coordinators	Glenn Davis, Kim Thomson

4 TC 03

Printed in Canada

PHOTO CREDITS

Chapter 1
page 1: Bob Semple; page 4:Dave Starrett; page 7: top left: PhotoDisc, top right; Dave Starrett, centre left: PhotoDisc, centre right: Corel, bottom left: Bob Semple, bottom right: PhotoDisc; page 13: Dave Starrett; page 14 top: PhotoDisc, bottom: Dave Starrett; page 16: Bob Semple; page 26: Bob Semple; page 27: Bob Semple; page 30: Bob Semple; page 31: Bob Semple

Chapter 2
page 55: PhotoDisc; page 59: Dick Hemingway; page 60: Corel; page 62 top: Corel, bottom: PhotoDisc; page 65 top: PhotoDisc, bottom: Corel; page 69: PhotoDisc; page 73: Corel

Chapter 3
page 95: PhotoDisc; page 102: Dick Hemingway; page 126: PhotoDisc; page 127: PhotoDisc; page 145: Images of New Brunswick; page 146: Bob Semple

Chapter 4
page 155: PhotoDisc; page 160: Canapress; page 162: PhotoDisc; page 169: Canapress; page 179: PhotoDisc; page 182: Dick Hemingway; page 190: PhotoDisc

Chapter 5
page 211: PhotoDisc; page 219: PhotoDisc; page 231: Canapress; page 240: PhotoDisc; page 243: PhotoDisc

Chapter 6
page 257: Dick Hemingway; page 258: Jeremy Jones; page 269: Dave Starrett; page 276: Dave Starrett

Chapter 7
page 305: PhotoDisc; page 306: W. Taufic/First Light; page 308: PhotoDisc; page 316: The Toronto Star

Contents

Chapter 1 Data Management

Introduction   1
1.1 Variables and Relationships   2
1.2 Measuring   7
1.3 Describing Data   14
1.4 Defining Data Spread   27
1.5 Large Distributions and the Normal Curve   33
1.6 Using Data to Predict   40
   Putting It Together   45
   Review   47
   Practice   52

Chapter 2 Networks and Matrices

Introduction   55
2.1 Creating and Traveling Network Graphs   56
2.2 Digraphs and Adjacency Matrices   63
2.3 Matrix Multiplication   74
   Putting It Together   84
   Review   88
   Practice   92

Chapter 3 Patterns, Relations, Equations, and Predictions

Introduction   95
3.1 Describing Patterns   96
3.2 Solving Problems by Solving Equations   104
3.3 Decision Making and Patterns   111
3.4 Predictions and Lines: $y = mx + b$ 117
3.5 More Patterns   130
3.6 Other Patterns   141
   Putting It Together   145
   Review   147
   Practice   152

Chapter 4 Modeling Functional Relationships

Introduction   155
4.1 Tables, Graphs, and Connections   156
4.2 Relations and Functions   163
4.3 Equipping Your Function Toolkit   174
4.4 Algebraic Models: Part 1   190
4.5 Algebraic Models: Part 2   199
Putting It Together   202
Review   205
Practice   208

Chapter 5 How Far? How Tall? How Steep?

Introduction   211
5.1 Ratios Based on Right Triangles   212
5.2 The Pythagorean Theorem   220
5.3 Square Roots and Their Properties   226
5.4 Defining Trigonometric Ratios   233
5.5 Applications of Trigonometry   241
Putting It Together   249
Review   251
Practice   255

Chapter 6 The Geometry of Packaging

Introduction   257
6.1 Examining Factors in Container Design   258
6.2 Regular Polygons   264
6.3 Surface Area   270
6.4 Economy of Design   275
6.5 Similarity and Size   283
6.6 Variations in Packaging   292
Putting It Together   296
Review   297
Practice   303

Chapter 7 Linear Programming

Introduction 305
7.1 Exploring an Optimization Problem 306
7.2 Exploring Possible Solutions 310
7.3 Connecting the Region and the Solution 319
Putting It Together 329
Review 331
Practice 334

Trigonometric Tables 336

Index 337

Chapter One
Data Management

Every day, people use numbers to help them make decisions. Numerical data are collected in sports, medicine, entertainment, and farming, to name just a few areas. For example, tree farmers might gather data about how quickly trees grow, to help them predict which trees can be sold in the coming year and what income they can expect.

In this chapter, you will look at effective ways to collect, organize, represent, display, and analyze data.

After successfully completing this chapter, you will be expected to:

1. Identify independent and dependent variables, and variables that must be controlled to determine cause and effect.

2. Take, calculate, and report measurements with appropriate precision and relate the precision to the number of significant digits.

3. Describe how data are distributed referring to measures of central tendency and dispersion.

4. Represent and interpret data displays such as stem-and-leaf plots, box-and-whisker plots, histograms, and scatter plots.

5. Use features of the normal curve to determine whether a measured value is realistic.

6. Construct lines of best fit for scatter plots and use them to make predictions.

1.1 Variables and Relationships

Tree farming is an important business in the Atlantic provinces. Christmas-tree farmers need to be able to predict how many trees of different sizes they will need each year in order to meet the demands of their customers.

A study of the factors that affect tree growth can help farmers better understand tree growth. This will help them make predictions about what sizes of trees will be ready for the next season.

Focus A: Cause-and-Effect Relationships

variable – any measured quantity that changes in an experiment or relationship

In any relationship, factors that can change are called **variables**. When a change to one variable causes a change in another variable, a *cause-and-effect* relationship exists between the variables.

A *mind map*, like this one, can be used to show cause-and-effect relationships.

– Note –
Diagrams of this type are often used to represent and examine cause-and-effect relationships. They are called mind maps. They show how the variables in a relationship relate to each other.

Mind Map of Factors that Affect Tree Growth

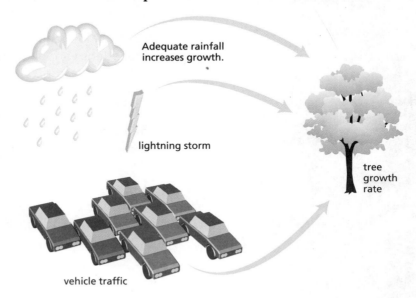

independent variable – a factor that affects another factor in an experiment or relationship

dependent variable – the factor that is affected by other factors in an experiment or relationship

- In each relationship, the cause is the factor that affects tree growth in some way. It is the **independent variable**.
- In each relationship, the effect is the amount of tree growth. It depends on the values of the other quantities. It is the **dependent variable**.

Focus Question

1. These are some factors that affect tree growth.
   - amount of rain or snow
   - temperature
   - soil nutrients

   (a) List other factors that might affect tree growth. Decide if the factor can be changed by the farmer.

   (b) Copy the mind map in Focus A.
   - Include any additional factors from part (a).
   - For each factor, include a brief description of how it affects tree growth.
   - Identify whether each factor can or cannot be changed by the farmer.

> **Did You Know?**
>
> In the Maritimes, and in New Brunswick in particular, the production of Christmas wreaths is a multi-million dollar industry. Newfoundland and Labrador has an abundance of balsam fir, which is the species of choice. The removal of tips for wreath production, if done correctly, has no measurable impact on the production of wood fibre from the trees.

Check Your Understanding

2. Suppose a farmer were growing corn.
   (a) What might the dependent variable be? List some possible independent variables.
   (b) Which variables could the farmer change?
   (c) Choose one independent variable from (b) and describe the cause-and-effect relationship between it and the dependent variable from (a).

3. Ski jumping is a popular event at the Winter Olympic Games. The winner is the person who jumps the greatest distance.
   (a) Explain why the distance jumped is the dependent variable.
   (b) List some of the factors that might affect the distance jumped.
   (c) List the factors in (b) that:
      (i) will be affected by the decisions jump designers make.
      (ii) will be affected by the decisions the skier makes.
      (iii) are not affected by anyone's decisions.
   (d) Why are the factors in (b) and (c) called independent variables?
   (e) Draw a mind map to show the relationship between each independent variable and the dependent variable.

1.1 Variables and Relationships

Focus B: Controlling Variables

In any real situation, like ski jumping or growing trees, there are many factors that can affect the final result. It is difficult to know which factor has the greatest effect.

Scientists solve this problem by carefully designing experiments that allow one variable to be studied independently from the rest.

[The variable to be studied (the controlled variable) is allowed to change, while all other variables are held constant.] The scientist can tell what effect changing that one variable has on the experiment.

The experiment is called a controlled experiment.

Suppose you were to measure the length of time needed for a spot of glue to dry.

- List the independent variables.
- What is the dependent variable?
- Design experiments in which only your independent variable is allowed to change. Carry out your experiment.
- What, if any, conclusions can you make?
- Did all independent variables affect the dependent variable?

controlled variable – any independent variable whose value is held constant during an experiment

controlled experiment – any experiment in which all but one independent variable are controlled

Think about...

Question 4
Why is it important to hold some variables constant in an experiment?

Focus Questions

4. Sugar dissolves in water. Think about the factors that could affect how long it takes to dissolve.

   (a) List some of the variables that might affect the time it takes for sugar to dissolve. Identify those that are controllable.

   (b) What do you think you can do to each independent variable in (a) to decrease dissolving time?

   (c) Describe how you would set up an experiment to determine how the mass of the sugar affects dissolving time. Identify the variables that would have to be controlled, or held constant, to be certain that the mass is the variable that is causing the effect on dissolving time.

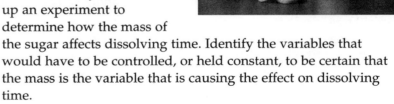

Check Your Understanding

5. Copy and complete the following table.

Description	Variables	Cause-and-Effect Relationship? Yes/No
(a) The length of a candle and the amount of time it has been burning.		
(b) The mark you get on a test and the time you spend studying.		
(c) The speed of a car and the distance from the nearest gas station.		
(d) The length of a movie and the admission price.		
(e) The age of a car and its current value.		
(f) The population of a community and its distance from the coast.		

6. An apprentice clockmaker is investigating the movement of a pendulum in a grandfather clock. The period (T) of a pendulum is the time it takes for the pendulum bob to make one complete swing (back and forth).
   (a) List several variables that you think might affect the period.
   (b) List the dependent and independent variables. Describe a set of experiments that you could conduct to find out which of the independent variables affect the period.

Pendulum
Question 6

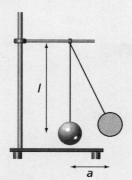

The length, l, is measured to the centre of the bob.

The amplitude, a, is the horizontal distance the bob moves to one side.

1.1 Variables and Relationships 5

Question 7

How might you predict the cost of a 2.5-m Scotch pine?

7. Vic wants to see how the price of various types of Christmas trees is set according to the height. He gathered the following data from some flyers.

White pine, 2 m, $24	Scotch pine, 2 m, $32
White pine, 2.4 m, $32	White pine, 2.2 m, $28
Scotch pine, 2.6 m, $56	Scotch pine, 2.4 m, $48
Douglas fir, 2.4 m, $25	Douglas fir, 2.2 m, $23
Douglas fir, 2 m, $21	Scotch pine, 2.2 m, $40

(a) Create a table to show how the price of a Scotch pine is set according to the height. Describe the relationship.

(b) For each of the other types of trees, create a table that shows how the price depends on the height.

8. The following table lists the cost of operating a 60-W light bulb for different amounts of time. Describe any patterns in the data.

Time used (h)	Cost ($)
1000	4.14
250	2.04
800	3.31
900	3.73

9. An experiment was conducted at a Christmas-tree farm to determine the effect of the amount of fertilizer on tree growth. The experiment was repeated in a greenhouse with different results. Which experiment do you think produced the more reliable results? Use what you know about controlling variables to explain why.

Chapter Project

Bouncing Balls to Different Heights

Have you ever played basketball and used a ball that did not bounce normally? How about a volleyball? Would you expect a volleyball to bounce normally?

Over the duration of this chapter, you will investigate the relationship between the type of ball and the height the ball bounces when dropped from a known height. You will be referring to the data, and analyzing it as you work through the chapter.

(a) List the dependent variables and the independent variables involved in the study of bounce height.

(b) Create a mind map.
  • Show how the independent variables interact.
  • Show how the dependent variables interact.

1.2 Measuring

A tree farmer might hire someone to measure the heights and diameters of his trees on a regular basis. The person who measures will be expected to be both accurate and precise. It is also important that the correct measuring tool is used.

Think about...

The Photographs

Different measuring tools are shown below. Which tools can be used to measure the same quantity?

For each tool, describe a situation in which it is the best choice for measuring.

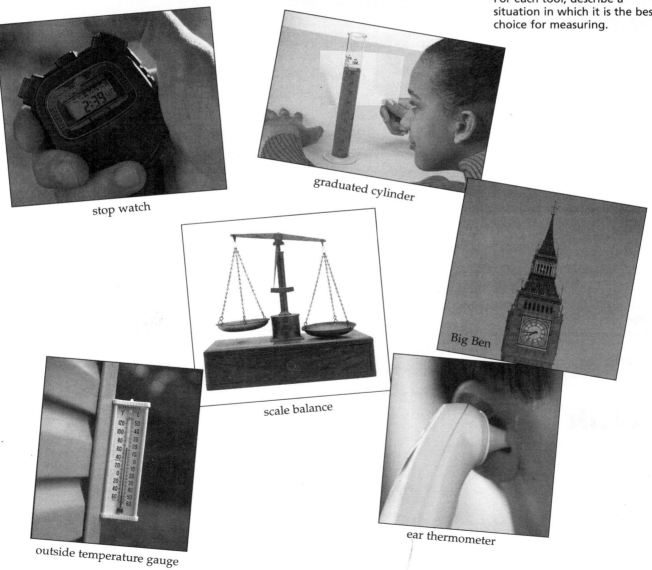

stop watch

graduated cylinder

Big Ben

scale balance

outside temperature gauge

ear thermometer

Accuracy and Precision

accuracy – the accuracy of a measurement indicates how close the recorded measurement is to the true value. It depends on the user's skill in using the tool. Other factors, such as temperature and humidity, can also influence the accuracy.

Think about...

Accuracy and Precision

Of the three measurements, which measurement shows:

- good accuracy and good precision?
- good precision but bad accuracy?
- bad precision and bad accuracy?

precision – the precision of a measurement tool is the smallest unit that can be measured with confidence using the tool. It depends on the fineness of the scale on the tool.

Think about...

Precision

If you are measuring a line that is exactly 10 cm long, why should you record the measurement as 10.0 cm instead of 10 cm?

ACCURACY

Three different people might use the same ruler to measure and get different lengths for this rectangle.

A One person places the ruler correctly and reads the scale properly as 6.8 cm.

B One person places the ruler and incorrectly records 7.8 cm.

C Another person does not look straight down at the scale when reading it and records 6.6 cm.

Even though all three people used the same ruler they read the measurement with different accuracy.

PRECISION

When you use a tool to measure a quantity, the precision of the measurement depends on how fine the scale divisions are. The length of the rectangle shown below is measured at:

- 6.8 cm using the centimetre scale; and
- 68.8 mm using the millimetre scale.

The finer scale allows you to record the measurement to a greater number of digits. This increases the precision of the measurement.

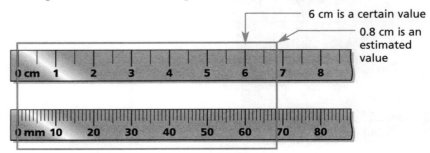

6 cm is a certain value

0.8 cm is an estimated value

8   Chapter 1 *Data Management*

COUNTING SIGNIFICANT DIGITS

When you do calculations involving measured values, your answer can only be as precise as the least precise measured value. **Significant digits** can help you make decisions regarding the least precise measured value.

The way significant digits are used is understood universally by scientists, engineers and anyone else who works with measurements.

Focus Questions

1. Suppose you measure a length using a ruler that is marked in decimetre divisions only. To what unit of measurement would you have confidence in your measurement?
2. Why is it important to use the same tool to measure length and width?
3. Explain why you should consider a reported measurement such as 100.0 m to be more precise than 100 m.
4. Why do you think the precision of a calculated measurement should be considered the same as the least precise measurement?

Investigation 1
Measurement Accuracy and Precision

Purpose
To explore precision and accuracy when measuring

Procedure

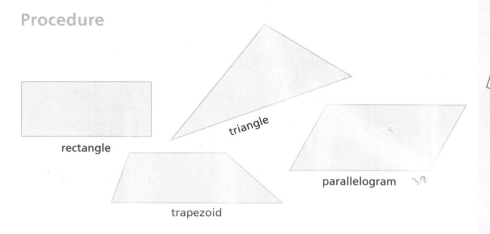

A. Use a centimetre ruler to measure the dimensions necessary to calculate the area of each polygon. Measure as accurately as you can. Record your measurements as precisely as you can. Calculate the areas.

$3.5 \times 1.5 = 5.25$

— Note —
Significant Digits
- All non-zero digits are significant. For example, the measurements 517, 51.7, and 5.17 all have 3 significant digits.

— Note —
- For a decimal number, any zeros that appear after the last non-zero digit (or between 2 non-zero digits) are significant. For example, 0.050 57, 5057, and 56.50 all have 4 significant digits.

— Note —
- For whole numbers, only zeros between two non-zero digits are significant. For example, 47, 470, and 4700 all have 2 significant digits.

— Note —
With a centimetre ruler, you could measure the length of a piece of paper as 27.9 cm. Using a millimetre ruler, you could report it as 279.1 mm, or 27.91 cm, which is a more precise measurement.

Think about...

Step C
Explain how the fineness of the scale on a measuring instrument affects the precision at which you can report a measurement.

Think about...

Question 8
What is meant when we say that the last digit in a measurement is estimated or uncertain?

Think about...

Accuracy and Precision
How accurate were your measurements? How precise were they?

B. Compare your two measurements and the calculated area for each polygon with another student's results. How do they compare?

C. Repeat steps A and B but this time use a millimetre ruler. Report each length in centimetres and each area in square centimetres. For example, report a measurement such as 82.2 mm as 8.22 cm.

D. Which area calculations were more precise, the ones found using the centimetre ruler or the millimetre ruler? Explain.

Investigation Questions

5. List some of the factors that might have caused you or your classmate to make inaccurate measurements.

6. Assume that two people measured the same rectangle accurately using the same tool. Will their measurements be exactly the same? Explain.

7. Suppose you had used a decimetre ruler in the Investigation. How would your results have compared to your results from Step B? Explain why.

8. If you use a centimetre ruler, you can confidently report a measurement to the nearest millimetre. Explain why.

Check Your Understanding

9. Record each measurement with the appropriate level of precision.

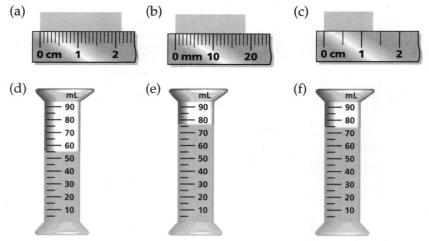

10 Chapter 1 *Data Management*

FOCUS D: Calculating with an Appropriate Level of Precision

When you calculate with measurements, your answer will not be any more precise than your least precise dimension.

Example 1
You are told that a room measures 6.2 m by 12.63 m. Find the perimeter of the room. Use appropriate precision in your answer.

Solution
Perimeter = 2 × (width + length)
Perimeter = 2 × (6.2 m + 12.63 m)

   (1 decimal place: (2 decimal places)
   less precise)

= 2 × 18.8 m
= 37.6 m

—Note—
When adding or subtracting, the right-most digit in the result should be in the same position as that in the least precise measurement.

Example 2
Suppose you used a centimetre ruler to measure the length of a rectangle and recorded 10.3 cm. A classmate measured the width using a millimetre ruler and recorded 52.3 mm. Report the area using appropriate precision.

Solution
$A = lw$ $l = 10.3$ cm = 103 mm $w = 52.3$ mm
$A = 103$ mm × 52.3 mm
$A = 5386.9$ mm$^2 \approx 5390$ mm^2

—Note—
When multiplying or dividing, the number of significant digits in the result is the same as that of the measure of the least number of significant digits.

Focus Questions

10. Explain why the area in Example 2 (5390 mm^2) was reported with no decimal places.
11. Repeat Example 2 above, but this time report the area in square centimetres. Explain what you did and why.

Check Your Understanding

12. Write your answers with the appropriate number of significant digits.
 (a) 3.7 + 1.6   (b) 3.75 − 1.6   (c) 3.75 × 1.125
 (d) 3.7 ÷ 1.25   (e) 370 × 1.875   (f) 3.78 ÷ 2.0

13. (a) Find the area of this triangle. Be as accurate as possible when measuring and report the area with the appropriate level of precision.

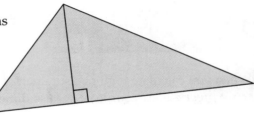

(b) Will your results likely be exactly the same as a classmate's? Explain.

(c) What might cause inaccuracies in the results?

14. Environment Canada stated yesterday that the temperature was 21.225°C. The temperature for today is shown on the thermometer.

(a) Read the thermometer as precisely as possible.

(b) Calculate the difference between yesterday's and today's temperatures. Write your answer with the appropriate precision.

(c) Suppose you were a weather forecaster for a radio station. What would you tell your listeners about the difference in the temperatures? Give reasons.

15. Calculate the area of each coloured region. The scales are in centimetres.

(a)   (b)   (c)

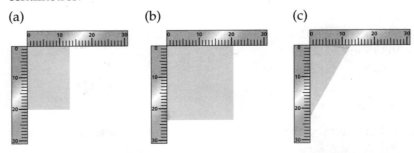

16. Each member of Fred's group measured the dimensions of their classroom.

Percy: 3 m × 5 m × 4 m

Wanda: 3.1 m × 5.2 m × 3.9 m

Rex: 310 cm × 535 cm × 405 cm

Ming: 3107 mm × 5270 mm × 4058 mm

(a) In whose measurements do you have the most confidence for further calculations?

(b) Find the volume of the classroom using each person's measurements. Which calculation would you use for further calculations? Why?

> **Did You Know?**
>
> Canada could be affected by global warming. Environment Canada predicts that the annual "average" temperature will increase by 2°C to 4°C in southern Canada. The increase will be higher in northern Canada.

17. (a) Measure and record the dimensions of your classroom.
    (b) Calculate the capacity of your classroom. Which number of significant digits will you use?

18. The label on a can of apple juice says it contains 1.36 L.
    (a) How many significant digits is this?
    (b) Would it be appropriate to label the capacity as 1.360 L? Explain.

19. Sally measured the length of a desk and reported it as 100 cm. Jennifer measured the same desk and reported it as 100.0 cm.
    (a) What might the different reported measurements indicate about the measurement tools they used?
    (b) Are the measurements equally precise? Explain.

20. A ball was bounced and the height of the first bounce was measured by five different students.

    Height of Ball Bounce

    | Student 1 | 75 cm |
    | Student 2 | 73.8851 cm |
    | Student 3 | 74.2 cm |
    | Student 4 | 74.4 cm |
    | Student 5 | 74.3 cm |

    (a) In which pieces of data would you have the most confidence? Why?
    (b) If you were to bounce the ball again, how high do you think it would bounce? Explain.

21. A child's height was measured five times and the values recorded as: 0.75 m, 75 cm, 75.2 cm, 750.3 mm, 75.38 cm

    Which measurement would you use to tell the class the child's height? Why?

Chapter Project

Bouncing Balls to Different Heights

Select a type of ball whose bounce height you would like to find.
- Drop the ball five times from a certain height and record the height of the bounce each time.
- Now, repeat the five drops for 11 different heights.
- Find the "average" bounce height for the ball.
- What did you do to make sure you reported bounce height with appropriate accuracy? Precision? Significant digits?

1.3 Describing Data

An athlete practising for a 400-metre race practises more than just running. She also practises starts. To compete successfully, she has to develop a consistently quick time out of the starter's blocks. Her coach times her to find her typical reaction time so that he can modify her training and plan a race strategy.

Investigation 2
Reaction Time

Purpose
To communicate your reaction time to an event

Procedure

A. Hold a 30 cm ruler at the top in a vertical position so that the 0 mark is at the bottom.
   - Have another person hold out one hand with thumb and forefinger on either side of the ruler at the 0 mark but not touching it.
   - Drop the ruler without warning the other person.
   - Record the position, in centimetres, on the ruler at which the other person grabs it.

B. Repeat step A at least 20 times for each person including yourself. (Keep your 20 pieces of data for use later in question 15.)

C. Find a single number that you feel best describes the position at which you will likely grab the ruler. This number describes your "typical" or **average** performance.

D. Compare your average number with those of your classmates. Discuss the different methods used to find the numbers.

Think about...

Step A
Explain how a linear measurement, such as 15.2 cm, is used to measure reaction time.

Step B
Why is it necessary to take more than one measurement for each person? Discuss why an individual's results might not be the same for each drop.

Step C
Why would an average based on more than 20 measurements be more reliable?

average – a number that is typical of a set of numbers

The most common averages are
- mean;
- median; and
- mode(s).

They are called "measures of central tendency."

Investigation Questions

1. Suppose you repeated the experiment 100 times and then found your average.
   (a) What do you think would happen to your reaction-time measurements?
   (b) How might your new average compare with your average based on 20 pieces of data?
   (c) How does increasing the amount of data affect your confidence in your calculated personal average?

2. Look at your set of reaction-time data and compare it with your average.
   (a) Select a piece of data that you would classify as "above average" and one you would classify as "below average." Explain your reasoning.
   (b) Are there any pieces of data that you feel should not have been included when you found your average? Explain.

Average—A Single Number Used to Describe or Represent a Set of Data

When using an average to describe a set of data, it is important to understand what it means. Suppose customers spend an average of 20 min in a fast-food restaurant. This does not mean that every customer spends exactly 20 min, but that most customers spend about 20 min at the restaurant.

It is important to keep in mind that, even with an average of 20 min, there are some people who spend 5 min or 2 h eating a burger and fries. These pieces of data, 5 min and 2 h, are called **outliers** because they are unusual values that "lie outside" most of the data. An outlier may be accurate but it is not typical.

outliers – values that are significantly different from the majority of a set of data

Averages are single numbers that can be used to describe data. It is important to choose the most useful average for analyzing your data.

A. The gas prices at six different stations were recorded and the following information was found.

   mean = 54.7¢ ✓
   median = 54.5¢ ✓
   mode = 54.5¢ ✓

   Which measure is the best? Why?

– Note –
Remember that the common averages are:
- mean;
- median; and
- mode.

1.3 Describing Data   15

B. Eight house prices were recorded and the following information was found. Which measure is the best? Why?

mean = $62 112.38
median = $55 200.00
mode = $50 000.00

C. The number of children in a family was recorded for 20 families and the following information was found. Which measure is the best? Why?

mean = 2.35
median = 2.5
mode = 3

Focus Questions

3. For each set of data, choose the most useful average. Then find it. Explain why you chose that average.
   (a) 2, 2, 12, 13, 14, 15, 16, 17, 18 (pencil lengths, cm)
   (b) 90, 100, 110, 120, 130, 140, 150, 160, 170 (duration of phone calls, s)
   (c) 5, 6, 6, 6, 7, 7, 8, 9, 12 (shoe sizes)

4. Look at your 20 pieces of data from Investigation 2. What average do you think is the best one to use to describe your typical reaction time? Explain why.

CHALLENGE yourself

Create one set of ten pieces of data which meets both of these conditions:

- The mean is double the median.
- The mode is double the mean.

Is there a way to describe the solution in general?

Focus F: Data Distribution—Stem-and-Leaf Plots

This table contains the average ruler-drop reaction times collected for a group of 40 students. It is not organized in any way. As a result, it is hard to see if there is any pattern in the distribution.

Group 1: Average Reaction-Time Data

26.0	26.3	23.5	28.6	23.5	15.8	17.2	26.2
16.4	17.7	27.5	17.5	20.6	22.7	14.1	11.2
20.8	16.5	16.2	28.9	10.7	15.6	13.8	26.4
19.2	28.8	16.4	11.1	18.5	28.8	15.9	16.5
27.2	17.2	23.3	10.2	15.2	18.0	28.5	29.1

A stem-and-leaf plot can be used to organize the same data so they are easier to analyze. For these data:

- This *stem* shows the tens and units digits.
- This *leaf* shows the tenths digit.
- Each data value is read by combining the stem and the leaf. For example, in this data the fastest reaction time is 10.2 and the slowest is 29.1.
- The data are in order from least to greatest.
- The range of the data can be calculated by finding the difference between the least and greatest values. The range of Group 1's data is 29.1 − 10.2 = 18.9.
- The "shape" of the plot gives an indication of how the data are distributed.

To construct a stem-and-leaf plot, divide each piece of data into two parts: a stem and a leaf. The last digit of each number is the leaf; the remaining digits comprise the stem. The data are then organized by grouping together data items that have common stems.

Focus Questions

5. Examine the stem-and-leaf plot in the margin.
   (a) List the five fastest reaction times, starting with the fastest.
   (b) List the five slowest reaction times, starting with the slowest.
6. (a) In the stem-and-leaf plot in the margin, it appears as if the data *cluster* in a few places. Where are the clusters?
   (b) What do these clusters tell you about the class reaction times?
   (c) Describe how the data are spread. What does the spread tell you about the class performance?

---

distribution – how all the data values in a set of values are spread

stem-and-leaf plot – an organization of data into categories based on place value

range – the difference between the least value and greatest value in a set of data

Stem-and-Leaf Plot

Stem	Leaf	Count
10	27	2
11	12	2
12		
13	8	1
14	1	1
15	2689	4
16	24455	5
17	2257	4
18	05	2
19	2	1
20	68	2
21		
22	7	1
23	355	3
24		
25		
26	0234	4
27	25	2
28	56889	5
29	1	1

— Note —
A stem-and-leaf plot
- organizes data in order of size
- can be used to find the range and all three measures of central tendency
- shows how data are distributed

1.3 Describing Data

Think about...

Question 7
Explain how a stem-and-leaf plot makes it easy to find the median of a set of data.

7. For Group 1's reaction-time data, the mean is 20.2.
   (a) Find the median.
   (b) Which measure of central tendency best represents this set of data? Explain why.

8. What does a stem-and-leaf plot show about the data that an average does not?

FOCUS G: Data Distribution—Box-and-Whisker Plots

box-and-whisker plot – a type of graph used to display data. It shows how data are dispersed around a median, but does not show specific items in the data.

A box-and-whisker plot can also be used to display data in order to see how they are distributed. This table shows ruler-drop reaction times for another group of 40 students. As you can see, the data are already in order from fastest to slowest reaction times.

Group 2: Average Reaction-Time Data

12.1	12.2	12.7	14.2	15.9	16.2	16.4	16.4
16.5	16.5	17.1	17.2	17.2	17.2	17.5	17.6
17.7	17.8	17.8	18.0	18.2	18.5	19.6	20.0
20.8	21.5	22.2	23.0	23.5	23.7	26.3	26.4
27.1	27.1	27.2	27.6	27.9	28.5	28.5	28.6

The following steps are used to create a box-and-whisker plot:

Step 1 Construct a number line and mark the **lower** and **upper extremes**, 12.1 and 28.6. The difference between the extremes is the range of the data.

lower and upper extremes – the least and greatest data values

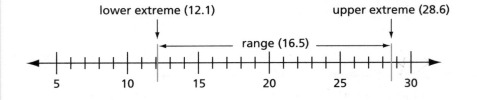

18 Chapter 1 *Data Management*

Step 2 Find the median of the data. Mark this value, 18.1, on the number line.

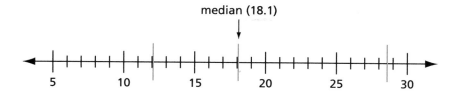

Step 3 Find the lower quartile. Mark this value on the number line. The median is found using the mean of 16.5 and 17.1, which is 16.8.

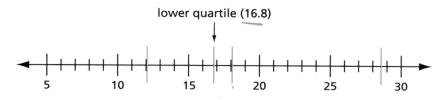

Step 4 Find the upper quartile. Mark this value on the number line. The median is found using the mean of 23.7 and 26.3, which is 25.0.

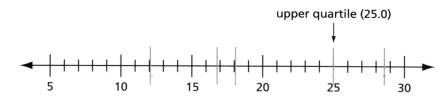

Step 5 Construct a box to show where the middle 50% of the data are located.

Group 2's Average Reaction Time

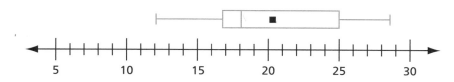

lower quartile – the median of the lower half of the data

upper quartile – the median of the upper half of the data

1.3 Describing Data

Focus Questions

9. Are the lower and upper extremes in a set of data always outliers? Explain.

10. What does a box-and-whisker plot tell you about the data that an average cannot?

11. One way to compare how sets of data are distributed is to compare box-and-whisker plots. Below are box-and-whisker plots for Group 1 (from Focus F) and Group 2.

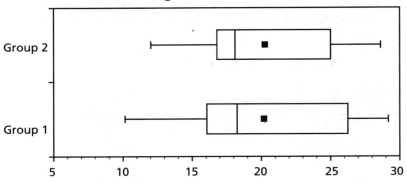

Average Reaction Times

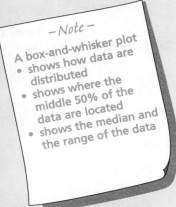

— Note —
A box-and-whisker plot
- shows how data are distributed
- shows where the middle 50% of the data are located
- shows the median and the range of the data

(a) Which set of data is more spread out?
(b) Which set of data has outliers furthest away from the centre?
(c) Which group had faster reaction times overall? Explain how you arrived at your decision.

12. Compare stem-and-leaf plots to box-and-whisker plots as ways to analyze data. What are some advantages and disadvantages of each?

Think about...

Question 12
When might a stem-and-leaf plot be a useful step in creating a box-and-whisker plot?

Check Your Understanding

13. Each box-and-whisker plot below was constructed using 40 pieces of data collected from a pair of students who had completed Investigation 2. For each, describe:

   (i) what you consider to be a typical reaction time.
   (ii) what reaction times you consider to be outliers.
   (iii) a range of values inside which the middle 50% of the data fall.

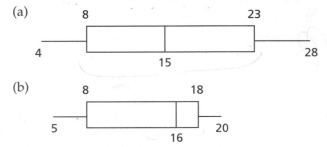

Did You Know?

American statistician John Tukey invented box-and-whisker plots in 1977. He wanted a quick way to visually represent a data distribution and to show how an individual score fits.

14. (a) The numbers below show the ages, in years, of 25 trees in a pile of firewood. The age is determined by counting the number of rings in the trunk of the trees. Construct a stem-and-leaf plot and then a box-and-whisker plot both by hand and by using a graphing calculator.

    14 14 15 15 18 19 24 24 25 27 28 28 28
    31 34 35 35 36 37 38 39 39 41 41 41

    (b) Describe what the plot tells you about the typical ages of these trees.

15. (a) Use 40 pieces of ruler-drop data from Investigation 2 to construct a stem-and-leaf plot and a box-and-whisker plot. Make your box-and-whisker plot using a graphing calculator.

    (b) Compare the box-and-whisker plot and the stem-and-leaf plot. Which is the more effective display? Why?

16. The number of super-deluxe hamburgers sold each day was recorded for a fast-food restaurant. The restaurant wanted to know on what percent of the days 50 or fewer burgers were sold so that they would not be over-staffed. They also wanted to know on what percent of the days 100 or more were sold so they would not be understaffed. Organize the data so that you can easily find the needed results for the restaurant.

55	127	76	23	17	84	20	50	49	60	30	65	66
44	57	99	75	42	110	45	87	66	54	107	51	100
28	59	66	82	77	63	39	52	95	23	81	46	79
34	99	61	98	73	118	22	89	36	101	55	63	61
53	28	89	49	66	52	117	31	45	111	85	71	54
36	74	52	80	70	67	94	72	11	58	69	122	70
88	73	48	100	98	89	73	108	14	53	65	17	

17. An experiment compares the time required for 25 athletes to run 100 m.

    Experiment A (time in seconds)

    Relax 2 minutes.
    Take no deep breaths. Then run.
    12.80 12.64 12.03 12.50 12.19
    13.27 12.35 12.67 12.66 12.37
    12.65 12.05 12.51 12.48 12.01
    12.95 12.52 10.96 12.63 12.46
    12.93 11.73 12.16 12.51 12.34

    Experiment B

    Relax 2 minutes.
    Take 10 deep breaths. Then run.
    11.80 12.18 12.04 11.88 11.89
    11.90 10.43 11.78 11.45 11.77
    11.43 11.60 12.03 11.91 12.62
    12.32 12.06 12.02 11.47 12.04
    11.57 11.71 11.16 12.29 11.89

CHALLENGE yourself

The following box-and-whisker plots describe the bowling scores of five players on a team.

- Discuss the spread of the data for each player.
- Suppose one player is needed to bowl one playoff game. Whom would you select? Why?

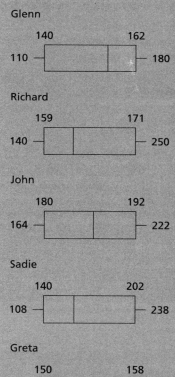

(a) Draw box-and-whisker plots for each set of data.

(b) What conclusions can you draw?

(c) Suppose you were to draw one box-and-whisker plot using both sets of data. What impact would it have on your conclusions?

18. An employer claims that 60% of employees drive to work. A sample of 20 employees revealed that 12 drive to work. Is the employer's claim likely to be correct? How could you check?

Data Distribution—Histograms

histogram – a graph in which the horizontal axis is a number line with values grouped in bins and the vertical axis shows the frequency of the data within each bin

bin – a grouping of data values

frequency table – a table that shows how often each data value, or group of values, occurs

The stem-and-leaf plot for Group 1's reaction times (from Focus F) gives a good indication of how the data are distributed. A **histogram** is another way to display data in graph form and is used when there are many pieces of continuous data.

To create a histogram, a **bin** size is chosen and then a **frequency table** is created. Below, a bin size of 5 has been used.

The bin 10–15 includes values that are 10 or greater but less than 15. Bin 15–20 includes values that are 15 or greater but less than 20, and so on.

Frequency Table

Bin	Frequency
10–15	6
15–20	16
20–25	6
25–30	12

The resulting histogram gives a picture of the distribution of the data not unlike that shown by the stem-and-leaf plot.

Did You Know?

Psychologists studying human visual ability have found that most people have difficulty distinguishing more than seven individual objects in a group of objects.

Histogram for Group 1

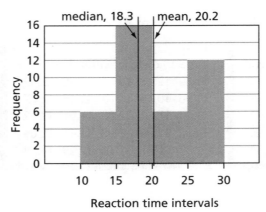

Think about...

Histograms

Redraw the histogram using bins of size 2, 4, and 6. Which gives a better picture of the distribution? Why?

Focus Questions

19. The following histogram represents the same data as that shown in the Focus but uses a bin size of 2.5. Copy it into your notebook.

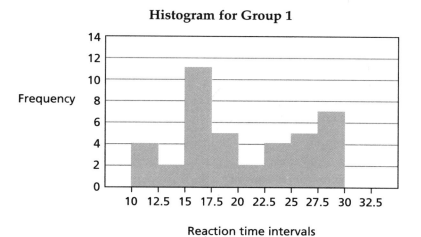

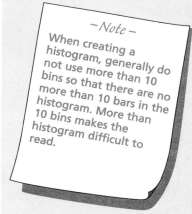

—Note—
When creating a histogram, generally do not use more than 10 bins so that there are no more than 10 bars in the histogram. More than 10 bins makes the histogram difficult to read.

(a) Use vertical lines to show the locations of the mean and median.

(b) Which bin size, 2.5 or 5, do you think gives a better picture of how the data are distributed? Explain.

(c) Predict what you think a histogram using a bin size of 10 will look like. Create one to confirm your prediction.

20. Compare the stem-and-leaf plot, the histogram, and the box-and-whisker plot for Group 1's set of data. What are some advantages and disadvantages of each type of graph?

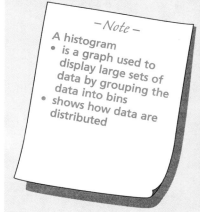

—Note—
A histogram
- is a graph used to display large sets of data by grouping the data into bins
- shows how data are distributed

Investigation 3
Comparing Reaction Times

Purpose
To communicate a description of data

Procedure

A. Use the same set of 40 reaction times used in question 15. Calculate the range of your data.

B. Copy and complete the following frequency table for your data.

Reaction Time	Frequency
0 to 5	
5 to 10	
10 to 15	
15 to 20	
20 to 25	

C. Use your frequency table to create a histogram.

D. Return to your original data from step A. Calculate the mean and median. Use vertical lines to indicate them on your histogram. Are the values close together?

E. Rate how your set of data is distributed using a 0 to 10 scale.

   0 = no spread (all the values are the same)
   3 = closely clustered together
   5 = moderately spread out, but more or less clustered together
   7 = spread out but only a few outliers
   10 = really spread out a lot

Investigation Questions

21. Explain why the range has no influence on your rating in step E.

22. What does the histogram show about the distribution of the data that an average cannot? Explain.

23. What would change most if you excluded outliers—the range, the distribution, or the average? Explain.

Think about...

Step A
- How do outliers affect the range of the data?
- Why does the range not give a good indication of how the data are distributed?

Step B
Explain how grouping data into bins makes them easier to manage.

Did You Know?
The TI-83 calculator can display two different types of box-and-whisker plots.

24. Compare the stem-and-leaf plot and the box-and-whisker plot from question 15 with the histogram. Compare how they show the distribution of the data.

Check Your Understanding

25. Compare the mean and median for the following ribbon lengths. Which measure of central tendency describes their average better? Give reasons.
    17.5 cm, 18.5 cm, 18.0 cm, 16.5 cm, 19.0 cm, 19.5 cm, 17.5 cm, 324.0 cm

26. After one week, the heights of 30 radish seed plants were measured. They are shown in the table.

    Radish Seedling Heights (mm)

5	25	22	32	10	25
21	27	29	39	37	18
10	36	11	33	48	16
24	36	45	38	7	26
15	37	17	22	25	13

    (a) Are there any values that you feel are outliers? Explain.
    (b) Find the mean and median.
    (c) Prepare a frequency table and histogram for all the data. Mark the mean and median on your histogram using vertical lines.
    (d) Use the scale in step E of Investigation 3 to rate the distribution.

27. Sketch histograms to illustrate each distribution rating.
    (a) 0 = no spread (all the values are the same)
    (b) 3 = closely clustered together
    (c) 5 = moderately spread out, but more or less clustered together in one area
    (d) 7 = spread out but only a few outliers
    (e) 10 = very spread out

28. Explain why a single piece of data is not an effective way to describe a set of data.

29. What factors should you consider about a set of data before making a decision based on those data? Explain why.

30. Isabelle is starting a tree farm. She has gathered data about trees grown by two different companies. She wants to grow the biggest fir tree in the shortest period of time. Which company should she choose? Use what you have learned about how to interpret data to help her make a good decision. Provide reasons for the decision.

   Firs 'R' Us (diameter in cm of 1-year-old trees at a height of 1 m)
   2.2 2.1 2.5 1.8 1.6 1.4 1.7 1.9 2.3 2.8 2.4 2.7

   Balsam City (diameter in cm of 1-year-old trees at a height of 1 m)
   1.6 1.7 1.9 2.9 1.4 1.8 1.6 1.4 2.1 2.6 2.4 2.4

31. Isabelle plans to choose a trucking firm to ship the trees. If a speedy delivery is important, which company should she choose? Explain.

   White Bay Trucking (trip times in hours)
   15.4 21.3 25.4 26.4 15.8 22.4 22.6 23.7 22.4 25.6

   Notre Dame Couriers (trip times in hours)
   26.6 26.7 28.1 31.2 34.5 35.6 36.6 38.4

   Central Transportation (trip times in hours)
   14.8 15.4 22.1 23.4 22.1 19.4 18.4 26.7 24.8

Chapter Project

Bouncing Balls to Different Heights

(a) What type of graph would best describe the relationship between height and bounce? Use this format to graph the relationship.

(b) When you calculated the "average" bounce height, which measure of "average" did you use? Do you want to change your "average" now? Why? If so, you should make the change now!

1.4 Defining Data Spread

In section 1.3, you discovered that an average alone was not enough to describe a set of data effectively. The description of a set of data should also include information about how the data values in that set are spread. The spread of a set of data includes its range and the variation of the data within the range. In this section, you will find out how variation in a set of data can be defined and described.

FOCUS 1: Standard Deviation: A Measure of Variation

The **dispersion** of a set of data includes its range and the amount of variation in the data. The range is only a measure of how spread out the extreme values are, so it does not provide any information about the variation within the data values themselves.

Mathematicians have developed a way to measure the variation in a set of data called the **standard deviation**. The standard deviation is a number that can be used to represent the variation in a set of data. It takes into account how far away each data value is from the mean. If most of the data are clustered around the mean, there is little variation and the standard deviation will be low. If the data are more spread out or there are a lot of different data values, meaning there is a lot of variation, the standard deviation will be greater.

> **dispersion** – a measure of the spread of data including the range of the data and the variation within the set of data
>
> **standard deviation** – a number that describes the spread within a set of data. It represents the average distance a random piece of data is likely to be located from the mean of the data.

Example
The manager of a Christmas-tree farm did a study on the time (in hours) needed to sell scotch pines versus white pines once the trees were cut down. The results are shown below.

Scotch pines: 2.15   7.34   9.14   9.50   10.3   10.5   10.7   12.1   21.7
White pines: 2.28   2.46   3.25   4.25   10.2   13.5   14.4   21.2   21.7

Compare the times for the two types of trees.

Solution

1. Is the time needed to sell one type of tree less than the other?
   Compare the averages.
   The mean time for scotch pines is 10.4 h.
   The mean time for white pines is also 10.4 h.
   The average times for both are the same.

Did You Know?

In some sports, such as figure skating and diving, the lowest and highest scores awarded are often dropped and only the remaining scores are totaled.

2. Is the range of time to sell one type of tree greater than another? Compare the ranges.

   Scotch pine's time has a range of 19.6 (2.15 to 21.7).

   White pine's time has a range of 19.4 (2.28 to 21.7).

3. Is there more variation for the time needed to sell one type of tree than the other?

   Create a histogram and calculate the standard deviation for each set.

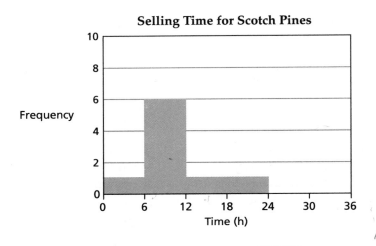

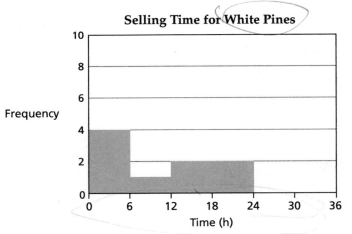

The method for manually calculating the standard deviation for white pines is shown below:

Step 1 Calculate the mean. It is 10.4.

Step 2 Calculate the distance each data value is from the mean. This is called the *deviation from the mean*. Square each deviation.

Think about...

Step 2

How would the standard deviation be affected if the deviations were used rather than the squares of the deviations?

White pine

Data Value	Deviation from the Mean (mean − data value)	Squared Deviation
2.28	10.4 − 2.28 = 8.12	65.9
2.46	10.4 − 2.46 = 7.94	63.0
3.25	10.4 − 3.25 = 7.15	51.1
4.25	10.4 − 4.25 = 6.15	37.8
10.2	10.4 − 10.2 = 0.2	0.04
13.5	10.4 − 13.5 = −3.10	9.61
14.4	10.4 − 14.4 = −4.00	16.0
21.2	10.4 − 21.2 = −10.8	116
21.7	10.4 − 21.7 = −11.3	128

Step 3 Find the mean of the squared deviations.

$$\frac{65.9 + 63.0 + 51.1 + 37.8 + 0.04 + 9.61 + 16.0 + 116 + 128}{9} = 54.2$$

Step 4 Find the square root of the mean of the squares from step 3.

$$\text{Standard Deviation} = \sqrt{54.2} = 7.36$$

A standard deviation of 7.36 for white pines indicates that a random piece of data taken from that set is, on average, 7.36 h, or a bit more than 7 h away from the mean. Most scientific calculators will find standard deviation.

Focus Questions

1. Compare the histograms for scotch pine and white pine. Do you think scotch pine will have a lower or greater standard deviation than white pine? Explain why. Higher because outlier

2. (a) Use the process shown in the example to find the standard deviation for scotch pine.
   (b) Based on the standard deviations, which set of times has more variation? Is this what you predicted from comparing the two histograms?

3. Suppose another type of tree, Douglas fir, produced a set of times with a greater range than white pine. Does that mean this set would automatically have a greater standard deviation? Explain. Support your explanation by using a sample set of nine times.

Think about...

Standard Deviation
What does each word, "standard" and "deviation," mean in the term "standard deviation"?

— Note —
- standard deviation is a measure of the variation in a data set
- the greater the standard deviation, the more the variation
- sets of data can have the same range and/or mean but different standard deviation values

Check Your Understanding

4. For each set of data below, create another set of data with the same number of data values and the same range, but with less variation. Explain what you did to create each set.

   (a) masses of rainbow trout landed during a fishing contest:

   1.2 kg, 5.1 kg, 1.1 kg, 0.8 kg, 1.3 kg, 1.1 kg, 0.9 kg, 1.3 kg, 1.1 kg

   (b) temperatures recorded at 1-h intervals during the day:

   $-3°, -2°, -1°, -1°, 0°, 1°, 2°, 3°, 2°, 1°, 0°, -2°$

5. For each set of data below, create another set of data with the same number of data values, the same amount of variation, and the same range, but with a different mean. Explain what you did to create each set.

   (a) speeds (km/h) of cars as recorded by a radar gun:

   95, 101, 97, 101, 105, 105, 106, 104, 98, 92, 101, 100, 96, 102

   (b) number of hours Jimmy watched TV each day for one week:

   1, 3, 0, 1, 9, 8, 3

6. Find the standard deviation of the set of data shown in each graph.

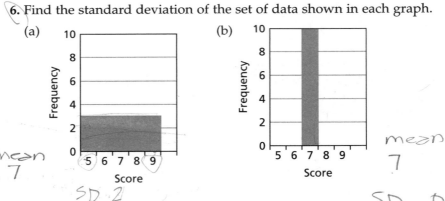

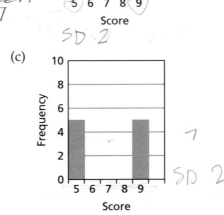

Did You Know?

- The number of Canadian households wired for cable television increased from 9 773 000 in 1992 to 10 249 000 in 1996.
- The total length of cable used increased from 168 000 km in 1992 to 190 000 km in 1996.

Source: Statistics Canada Website.

7. Quality control is a process used by companies to ensure that the amount of variation in their products is as small as possible. A glass manufacturer has three plants in three different locations. A batch of nine jars of the same design was randomly selected from each plant and then tested for the number of bubbles found within the glass. Which plant has the best performance? Explain how you know.

   Number of Bubbles in Glass Jars
   Plant 1   5, 6, 7, 8, 9, 10, 11, 12, 13
   Plant 2   5, 9, 9, 9, 9, 9, 9, 9, 13
   Plant 3   5, 7, 7, 7, 9, 11, 11, 11, 13

8. The diameters in centimetres for a group of 5-year-old black spruce trees grown in a forest are shown below.

   Group 1 Forest Environment (tree diameters in cm)

13.2	13.9	14.3	14.7	15.3	15.5
16.4	16.9	16.9	17.0	17.9	18.0
18.7	19.1	19.2	19.3	19.5	20.0
20.2	20.3	20.7	21.0	22.2	22.6
22.8	23.1	24.2	25.1	25.8	

   The chart below lists the diameters for another group of 5-year-old black spruce trees that were grown in a tree nursery.

   Group 2 Nursery Environment (tree diameters in cm)

19.3	19.3	19.8	20.4	20.8	20.9
21.5	21.7	22.6	22.6	22.7	22.8
22.9	23.1	23.2	23.4	23.5	23.6
23.7	23.7	23.8	23.9	24.0	24.1
25.2	25.3	25.4	25.9		

   (a) Use histograms and box-and-whisker plots to compare the distribution of the diameters in the two groups of trees. Which has greater variation? Explain why.

   (b) Use standard deviation to compare the variation in the two sets of data.

   (c) Did your findings in (b) confirm what you observed in (a)? Explain.

Did You Know?

In 1995, the following areas (in hectares) were planted with new tree seedlings:

Newfoundland   3453
P.E.I   837
Nova Scotia   7186
New Brunswick 16 162

Source: Statistics Canada Website.

Think about...

Question 9
If you adjust values in a set of data and, as a result, change the mean, does the standard deviation change? Why or why not?

Question 10
Explain, in your own words, why comparing the standard deviations of sets of data may be better than comparing their graphs.

CHALLENGE yourself

Create a set of 10 pieces of data for which the standard deviation is double the mean. Is this possible if all the data are clustered around the mean? Is it possible if they are not?

---

9. These two sets of data have the same range but different means. How would you expect their standard deviations to compare? Calculate to check.

   Set 1:   45, 50, 60, 100

   Set 2:   145, 150, 160, 200

10. Without calculating, match each standard deviation with the appropriate histogram. Explain your choice in each case.

    Standard Deviations: 0.50   1.16   2.87   3.67

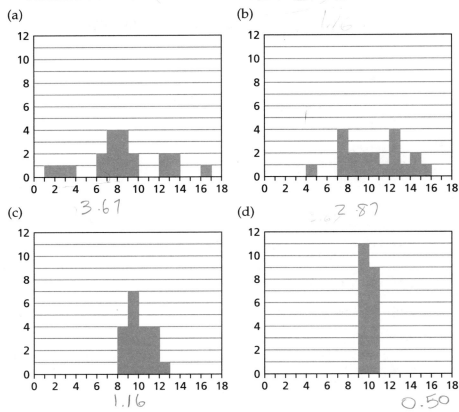

(a) 3.67   (b) 1.16   (c) 1.16   (d) 2.87   0.50

11. (a) Create three data sets meeting the following conditions:
    - There are at least five pieces of data.
    - The mean is 20 for each set.
    - The standard deviations are quite different for each set. (Calculate these to check.)

    (b) Construct a histogram for each set. How do the shapes of the histograms relate to the standard deviations?

12. Select at least one set of data from question 5. Change any of the data values except the two extreme values, keeping the same range and mean, but changing the standard deviation. Explain how you did this.

32   Chapter 1   *Data Management*

1.5 Large Distributions and the Normal Curve

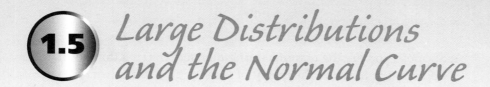

If you were collecting data to determine a typical height range for Grade 10 girls, you would feel more confident if your results were based on a large and random sample. In this section, you will investigate what happens to the averages and the distribution of a set of data as its size increases. This will give you a better understanding of why large, randomly selected samples are used in research.

Investigation 4
Comparing the Shape of Different-Sized Sets of Data

Purpose
To compare the distribution of a small set of data values to the distribution of a larger set.

Procedure

A. Create a histogram using a bin width of 10 for this set of height measurements (in centimetres) taken from a **random** sample of 19 Grade 10 girls.

Set 1: 150, 156, 158, 159, 165, 165, 169, 170, 170, 172, 173, 174, 175, 176, 185, 186, 186, 192, 193

B. Repeat Step A for this set of height measurements taken from a sample of 89 Grade 10 girls.

Set 2: 150, 152, 153, 154, 154, 155, 156, 157, 157, 160, 160, 161, 161, 162, 164, 164, 165, 165, 166, 166, 166, 167, 167, 168, 168, 168, 168, 169, 169, 169, 169, 170, 170, 170, 170, 170, 171, 171, 171, 171, 171, 172, 173, 174, 174, 174, 174, 174, 174, 174, 175, 176, 176, 176, 176, 177, 178, 178, 178, 179, 179, 180, 181, 182, 182, 182, 182, 183, 183, 185, 185, 186, 186, 186, 186, 186, 187, 187, 188, 188, 189, 192, 193, 193, 194, 194, 195, 195, 195

C. Create a **frequency polygon** for each histogram.

D. Find the mean, median and mode for each set of data. Draw vertical lines on the graph at each value.

E. Describe
- how the shape of Set 1 compares to the shape of Set 2; and
- how the locations of the averages in both sets compare.

random – when a data value is chosen at random from a set of data values, each possible data value has the same chance of being selected

frequency polygon – the shape that is formed when the centres of the tops of the bars of a histogram are joined by straight lines

frequency polygon

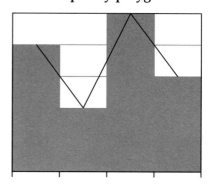

Think about...

Step E
How do you think the shape of a frequency polygon based on 10 data values would compare to this one? What about a frequency polygon based on 1000 pieces of data?

Investigation Question

1. Suppose you are the manufacturing manager for a clothing outlet. You want to measure the height of teenage girls to find an "average" height. Why would you expect:

   (a) a larger set of height measurements to have more values in the middle and fewer at the extremes than a smaller set of data?

   (b) the mean, median, and mode to be closer to each other in a set of 200 measurements than they would be in a smaller set of data?

The Effect of Sample Size on the Shape of a Frequency Polygon

We know from Investigation 4 that the size of the data set can affect the look of the frequency polygon describing the data. If there is a large number of data values, such as height measurements, the likelihood is greater that:

- the data are symmetrical around the middle;
- the three measures of central tendency are close together and in the middle;
- most of the data are clustered in the middle; and
- there are few extreme values.

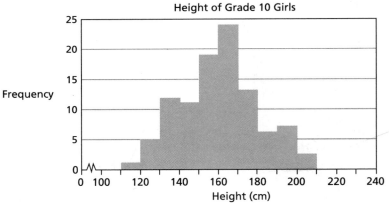

For example, the frequency polygon shows a frequency distribution for 100 height measurements of Grade 10 girls. You could use a graphing calculator to find histograms that have the following:

- the mean height is 160 cm;
- the standard deviation is 20 cm; and
- the mean, median, and mode are all the same.

The Data

Suppose you had 1000 height measurements of teenage girls. What shape do you think the graph will have? Why?

Focus Question

2. (a) Roll two dice 50 times. Find the sum of the pair of dice each time.
   (b) Construct a histogram for the data in (a). Describe the shape.
   (c) Ask another student for his or her data. Add it to yours and create another histogram. Can you now predict what shape a frequency polygon will have for 1000 pieces of data?
   (d) Construct a frequency polygon for 1000 pieces of data. How accurate was your prediction in (c)?

Check Your Understanding

3. Look at the frequency polygon below. Do you think the data were more likely collected from 20 people or 200 people? Explain why.

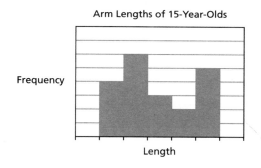

4. A trivia quiz, with a total possible score of 100, was given to all the Grade 10 students in a school. A random sample of 50 scores was selected:

   12 16 18 22 23 24 25 27 29 30 30 30 32 32 33 35 37 39 41 41
   42 43 44 45 45 45 45 48 48 49 49 50 50 50 50 56 57 58 58 61
   62 63 66 67 70 77 77 79 81 82

   (a) Create a histogram and frequency polygon for the data. Use a bin width of 10 for your histogram. Find the mean, median, and mode and locate them on the histogram.
   (b) Describe the shape of the frequency polygon.
   (c) Sketch what you think the polygon might have looked like if more data had been collected. Locate where you predict the averages would be on your polygon.
   (d) Find the standard deviation for the data.
   (e) Find the percent of data within one standard deviation of the mean and within two standard deviations.

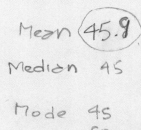

Normal Distribution

normal distribution –
a distribution
- in which the mean, median, and mode are the same and in the middle;
- such that the data are distributed symmetrically about the middle; and
- that has a graph that appears as a continuous bell-shaped curve.

When very large amounts of certain kinds of data are collected randomly and displayed in a histogram, the frequency polygon appears as a bell-shaped curve called a *normal curve*. The graph that follows shows a special bell-shaped distribution known as the normal distribution.

Normal Distribution Graph

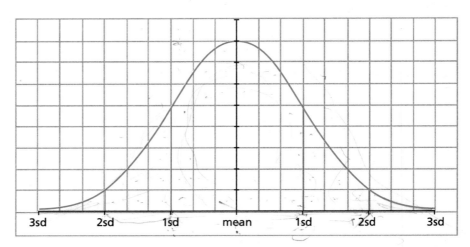

Examples of data that might be distributed like this are heights of 15-year-olds and pitching speeds of major league pitchers.

Some kinds of data will not be normally distributed no matter how large or random the sample might be. For example, if you roll a die many times and record the frequency of each possible outcome (1 to 6), the distribution would not be a normal distribution.

For a normal distribution,

- The mean, median, and mode are all the same value and are in the middle of the graph.
- The graph of the data is symmetrical about the middle.
- The ends of the curve get closer and closer to a frequency of 0.
- Find the number of squares within 1 standard deviation of the mean.
- Find the number of squares within 2 standard deviations of the mean.
- What percent of all the squares under the curve are within 1 standard deviation of the mean? Within 2 standard deviations?

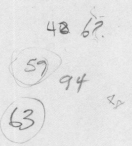

- Outliers fall outside of 2 standard deviations of the mean.

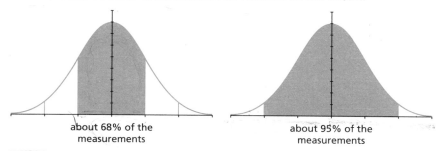

Think about...

Normal Distribution and Standard Deviation

Explain what is meant by
- within 1 standard deviation of the mean.
- within 2 standard deviations of the mean.

Focus Questions

5. Suppose 960 students took a math-achievement test out of 175.
   - The results were normally distributed.
   - The mean was 115.
   - The standard deviation was 25.

   (a) Within what range of scores would 68% of the students typically fall? Explain.

   (b) Within what range would most students fall? Explain.

6. Your friend claimed a score of 170 in the math test described in Question 5. Would you believe him? Explain why or why not.

7. A Grade 3 test was given to 1000 Grade 10 students, chosen at random. Would you expect a normal distribution? Explain.

8. Refer to your ruler-drop data.
   - Find 1 standard deviation from the mean.
   - What percent of your data fell within 1 standard deviation of the mean? Within 2 standard deviations?

— Note —
- a normal distribution typically occurs when the *sample* size is very large and the *data* are collected randomly.
- the standard deviation can be used to identify a range of typical values and to identify outliers

Think about...

Question 6

What percent of data values in a set of data distributed normally would be considered outliers?

Check Your Understanding

9. In a certain class, the mean number of hours of sleep on Friday night was 9 h. The standard deviation was 1.2 h. The data are normally distributed. Fill in the blanks to make these statements true.

   (a) 68% of the students slept between ___ h and ___ h Friday night.

   (b) 95% of the students slept between ___ h and ___ h Friday night.

   (c) It is very unlikely that a student in that class slept ___ h Friday night. Explain why.

1.5 Large Distributions and the Normal Curve

10. The weekend sleeping times for four classes are listed. Each set of data is normally distributed.

Class	Mean Sleeping Time	Standard Deviation
1	9 h	1.5 h
2	10 h	2.0 h
3	8 h	0.8 h
4	9 h	1.3 h

(a) Describe at least one unusual sleeping time in Class 1. Explain how you know it is unusual.

(b) A student slept 12 h Friday night. Which class is she most likely from? Why?

(c) A student slept 8 h Friday night. Why is it hard to tell which class he comes from?

11. From a questionnaire, Jared recorded everyone's height in his high school. In a separate chart, he recorded the heights of all the members of the senior basketball team.

(a) Which table, the one for the whole school or the one for the basketball team, will show the greater range? Why?

(b) Which table will have the greater standard deviation? Why?

(c) If the data were graphed, which frequency polygon would look more like a normal distribution? Why?

12. A survey of 1000 vacationing families produced the following information.
- The mean distance traveled was 700 km.
- The standard deviation was 40 km.

(a) Sketch a graph of how the frequency polygon might look.

(b) Suppose your family traveled 275 km. Would this value be considered an outlier in the survey distribution? Explain.

(c) Using the survey information, would you consider your family "typical" vacation travelers?

13. A restaurant owner studied the amount of time that each customer spent eating. Her restaurant seats 162 people and is usually full. She conducted the study over a 12-month period. She found that the mean time was 56 min and the standard deviation was 8 min.

(a) Would you expect this data to have a normal distribution? Explain.

(b) Would you consider it unusual for a customer to stay for 1 h 15 min? Explain.

Did You Know?

Some fast-food restaurants have uncomfortable seats so that customers will not stay too long. This means that more customers can be seated in a given period of time.

14. The Pan-African Film and Television Festival was first held in the winter of 1969 to spotlight African-made films. Countries outside Africa also participate.

Year	1969	1970	1972	1973	1976	1979	1981	1983
Number of films	23	40	36	51	75	78	78	69
Number of African countries	5	9	18	23	17	16	16	25
Total number of countries	9	9	23	29	26	26	27	37

In what year was there an unusual number of films spotlighted? Explain.

15. Refer to question 14.
    (a) Do you think that an unusual number of African countries participated? Explain.
    (b) Do you think that an unusual number of countries participated in total? Explain.
    (c) Has your understanding of "unusual" remained consistent in both questions 13 and 14? Did it need to change?

Chapter Project

Bouncing Balls to Different Heights

(a) Using all of the bounce heights you found, find the mean and standard deviation.

(b) Describe a "typical" bounce height for your ball.

(c) Calculate the typical bounce-height range for your ball.

(d) Could you tell by a "bounce height" reported by someone in your class whether your ball was being used?

1.6 Using Data to Predict

You have tried a variety of different methods of analyzing data. In this section, a scatter plot graph will be used to determine the relationship between two variables. Once a relationship has been established, the graph can be used to make predictions.

Scatter Plots and Lines of Best Fit

Age (years)	Mean Height (cm)
1	32
2	45
3	57
4	60
5	66
6	80
7	95
8	99

The heights of 1000 trees were measured each year over a 7-year period, from age 1 to age 8. The mean height at each age was calculated and then graphed.

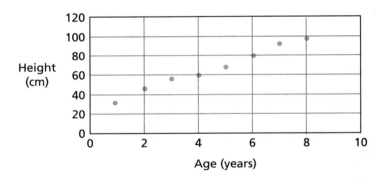

scatter plot – a graph of plotted ordered pairs that represents the relationship between two variables

The resulting graph is called a scatter plot. A scatter plot displays ordered pairs by using coordinates; for example, the first point (1,32) represents the relationship between 1 year and 32 cm.

- The independent variable (in this case, age) is plotted on the horizontal axis.
- The dependent variable (in this case, height) is plotted on the vertical axis.

As the name implies, the data points seem scattered on the graph because they are not connected. However, points often show a relationship, or pattern, that can be identified. In this graph, the points seem to form a straight line, which means they have a linear relationship.

line of best fit – line used in scatter plots to show a trend in data

A line of best fit drawn on a scatter plot can be used to show the relationship between the two variables.

40   Chapter 1 *Data Management*

Here is the same scatter plot with a line of best fit added.

- The line is not drawn through any particular points.
- It is drawn through, or close to, as many points as possible.

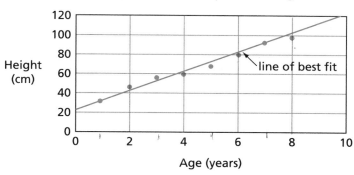

The arrangement of the points on the scatter plot and the corresponding line of best fit suggests a cause-and-effect relationship between the two variables, that is, as the tree gets older, the height increases. This indicates a relationship between age of tree and height.

Focus Questions

1. Examine the line of best fit.
   (a) Describe the relationship between the line and the eight data points.
   (b) To what age do you think it would be reasonable to extend the line? Explain your answer.

2. Suppose you could not find a line of best fit for a set of data points. What might this mean about the relationship between the two variables?   no cause & effect

Investigation 5
Predicting Writing Speed

Purpose
To graph recorded data and look for patterns in order to make predictions

Procedure
A. Find out how many times you can write the word "RED" completely and correctly in exactly 15 s.

> *- Note -*
> - scatter plots are graphs of ordered pairs used to determine the type of relationship between two variables
> - a line of best fit on a scatter plot approximates a linear relationship between the two variables

Think about...

Step B
Why is the median the best average to use here?

Step D
What other independent variables might influence the number of words you could write in 15 s?

Step F
Compare your line of best fit with that of others in your class. Is it reasonable to assume everyone has exactly the same line? Explain.

Steps G and H
Are all predictions based on a scatter plot and its line of best fit equally valid? Explain.

interpolate – make a prediction that falls between given pieces of data

extrapolate – make a prediction that is outside the range of given values

B. Repeat step A ten more times to collect a total of 11 data values. Find the median number of times.

C. Repeat Steps A and B using each of these longer words: GREEN, MAGENTA, TURQUOISE, and ULTRAMARINE.

D. Prepare a scatter plot that shows the relationship between the number of letters in a word and the number of times you can write it in 15 s. Which is the independent variable and which is the dependent variable?

E. Describe the pattern in the scatter plot. What does this tell you about the relationship between word length and the number of times you can write it?

F. Position a ruler on your graph to best represent the pattern you identified in step E. Draw the line of best fit. Describe what you did to locate the line.

G. Use your line to predict (*interpolate*) how many 6-letter words you could write in 15 s. How confident are you that your prediction is reasonable? Explain.

H. Use your line to predict (*extrapolate*) how many 2-letter and 12-letter words you could write in 15 s. How confident are you that your predictions are accurate? Explain.

Investigation Questions

3. Suppose someone told you they wrote the word "GREEN" 20 times in 15 s. Would you believe them?

4. Are there any reasonable limits on the predictions you could make about writing speed? Explain.

5. (a) Did your line of best fit pass through any of your data points? Do you think it should?

    (b) Write a paragraph explaining your answers to these investigation questions. Make sure you explain how you drew a line of best fit on your scatter plot. Include diagrams to support your explanation.

Check Your Understanding

6. Students were studying the relationship between slap-shot speed and goals scored in a hockey game. They matched the speeds against the scoring records from the past season for 13 players.

Shot speed (km/h)	80	100	93	138	115	82	134	95	145	91	120	108	121
Number of goals	6	20	10	58	35	5	42	15	50	14	35	30	15

(a) Construct a scatter plot. Draw a line of best fit.

(b) Use the graph.
  (i) Predict how many goals a player would score if he had a slap-shot speed of 110 km/h.
  (ii) A player scores 60 goals in one season. Predict her slap-shot speed in the playoffs.

(c) How much confidence should you place in your predictions in part (b)? Explain.

7. Statistics Canada produces reports based on Canadian data. The following chart shows how the average life expectancy has changed every 10 years since 1920.

(a) Graph the male and female data separately but on the same scatter plot.

(b) Use the graphs to predict the average life expectancies for Canadian males and females in the year 2020.

(c) What would the expected difference be?

(d) Discuss your confidence in your predicted results.

Life Expectancy in Years

Year	Male	Female
1920	58.84	60.60
1930	60.00	62.06
1940	63.04	66.31
1950	66.40	70.90
1960	68.44	74.26
1970	69.40	76.45
1980	71.88	79.06
1990	74.61	80.97

(Source: CANSIM database)

Technology

Throughout these questions you can use technology to find the line of best fit. You will explore more about the line of best fit in Chapter 4.

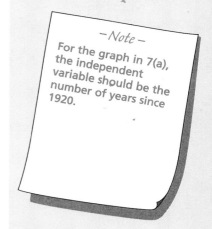

− Note −

For the graph in 7(a), the independent variable should be the number of years since 1920.

CHALLENGE yourself

Choose one of the topics below to research:

- the number of professional hockey players in North America since 1960
- sale of paperback books in Canada since 1950
- sales of computer games in Canada since 1985
- a similar topic of your own choice

Find data on this topic on the Internet or in reference books. Prepare a scatter plot. Predict the appropriate number for the year 2050. How much confidence do you have in your prediction?

8. Think about the graphs that you have created and seen in this section. What factors affect your confidence in extrapolated predictions? Use an example to help explain why.

9. In this chapter, you have used a variety of graphs to examine data.
   (a) How is a scatter plot different from the other graphs you have used?
   (b) Describe a situation in which you would use a histogram to analyze data. Explain why.
   (c) Describe a situation in which you would use a scatter plot. Explain why.

10. Refer to Focus B in Section 1.1 on page 4.
    - Draw a scatter plot of your data.
    - What conclusions can you draw regarding your data?

Chapter Project

Bouncing Balls to Different Heights

Suppose you were to drop your ball off of the roof of your school. How high do you think it would bounce? Draw a graph of your prediction to show others in the class.

PUTTING IT TOGETHER

CASE STUDY 1

Your local board of education is planning to provide schools with an information program dealing with health issues for teenagers. There are two programs from which the board must choose. One of the programs provides a set of printed information pamphlets. The other program is a videotaped presentation.

The school board's research department conducted a study of each program's effectiveness. The researchers selected two groups of students. One group used the pamphlet-based program and the other used the videotape. The students then completed a questionnaire designed to test their awareness and understanding of the issues. The student scores on the questionnaire appear in the tables in the margin.

(a) Construct frequency tables for both groups.
(b) Construct histograms for both groups.
(c) If the data are normally distributed, what conclusions could you make about the results?
(d) Do the data for each group appear to be normally distributed? Explain.
(e) Do the data clearly indicate that one program is more effective than the other? Use appropriate graphs and calculations to justify your conclusion. [Hint: Use measures of central tendency and standard deviation.]

Print Program

117	119	141	137
137	132	138	135
140	141	101	108
113	117	114	122
133	138	127	146
138	130	125	125
133	137	135	114
138	140	141	146

Video Program

145	161	164	143
164	151	159	166
158	172	167	140
161	139	143	167
134	156	151	164
164	153	172	163
167	140	148	160
163	163	166	127

EXTENSION 1

At a music festival, students were marked out of 100 by several judges and then the marks were averaged, as shown.

Junior piano - 85
Intermediate piano - 83
Senior strings - 87
Intermediate brass - 86
Intermediate strings - 84

Junior strings - 84
Junior brass - 88
Senior piano - 86
Senior brass - 88

(a) If you reorganize the data as shown to the right, you can "control" one of the independent variables in order to study the effect the other variable has on the dependent variable.
   • What is the dependent variable?
   • What is the independent variable being studied?
   • What is the independent variable being controlled?

Piano

Level	Score
Junior	85
Intermediate	83
Senior	86

Strings

Level	Score
Junior	84
Intermediate	84
Senior	87

Brass

Level	Score
Junior	88
Intermediate	86
Senior	88

(b) Describe how the independent variable seems to affect the dependent variable.

(c) Rearrange the data so you can see how the type of instrument affects the score.

(d) Describe how the independent variables changed from the situation in (a) to the situation in (c).

(e) Describe how the type of instrument seems to affect the score.

CASE STUDY 2

Try this experiment.

- Use a scrap piece of cloth and some thread to construct a parachute.
- Use a pen, nut, or bolt as a weight.
- Conduct an experiment to find the time it takes your parachute and weight to fall to the ground.
- Start with a height of about 1.5 m and then increase the height gradually.
- Collect as much data as you can.

(a) Use your data to construct a scatter plot.

(b) Construct the line of best fit.

(c) Predict the time needed for the parachute to drop 5 m.

EXTENSION 2

Repeat the experiment that you did in Case Study 2. This time, cut a small hole (about 1 cm in diameter) in the top of your parachute.

(a) Discuss whether cutting the hole made a significant difference in the drop time for the parachute. Use appropriate graphs and your knowledge of measures of central tendency and standard deviation to justify your conclusion.

(b) List as many possible independent variables as you can that could have an influence on the drop time. Choose two of these and explain how to design an experiment that would determine their effect on drop time.

REVIEW

Key Terms

	page
accuracy	8
bin	22
box-and-whisker plot	18
controlled variables	4
dependent variable	2
dispersion	27
distribution	17
extrapolate	42
frequency polygon	33
frequency table	22
histogram	22
independent variable	2
interpolate	42
line of best fit	40
lower extreme	18
normal distribution	36
outlier	15
precision	8
quartile	19
range	17
scatter plot	40
significant digits	9
standard deviation	27
stem-and-leaf plot	17
upper extreme	18

You Will Be Expected To

- read and interpret data values.
- record measurements with an appropriate level of precision.
- calculate with measured quantities and record the result with an appropriate level of precision.
- identify the correct number of significant digits in any number.
- describe data in terms of its location (measure of central tendency) and its spread, or dispersion (variation and range).
- interpret and create stem-and-leaf plots, box-and-whisker plots, frequency tables, and histograms.
- calculate and use standard deviation to compare the variation in data sets.
- list some features of normal distribution.
- explain the concepts of independent, dependent, and controlled variables.
- construct lines of best fit from scatter plots and use them to make predictions.

Summary of Key Concepts

1.2 Measuring

The **accuracy** of a measurement depends on the person's ability to use the instrument correctly. The **precision** of a measurement is the number of digits to which the measured value can be determined.

Example
Record the two measurements indicated in the diagram. The scales are in centimetres.

Solution
The horizontal length is 29.5 cm.
The vertical length is 23.5 cm.

How accurate your measurements are depends on whether or not you read the scales correctly.

The precision of your instrument depends on the fact that you can read these scales precisely to one decimal place.

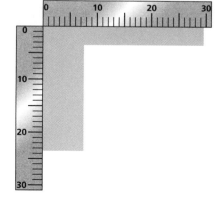

SIGNIFICANT DIGITS AND PRECISION

You can compare the precision of two measurements of the same thing by comparing the number of significant digits; for example, 3.45 cm is more precise than 3.5 cm because 3.45 cm has more significant digits.

You can determine the number of significant digits in a measurement using the following guidelines:

- All non-zero digits are significant. For example, 467, 46.7, and 4.67 all have three significant digits.
- For a decimal number, any zeros that appear after the last non-zero digit or between two non-zero digits are significant. Any zeros that appear before the first non-zero digit are not significant. For example, 0.040 67, 4067, and 46.70 all have four significant digits.
- For a whole number, any zeros that appear between two non-zero digits are significant. Unfortunately, you cannot tell whether any zeros after the last non-zero digit are significant so you should assume they are not unless told otherwise. For example, 47, 470, and 4700 all have two significant digits.

> *—Note—*
> When adding or subtracting, the right-most digit in the result should be in the same position as that in the least precise measurement.

> *—Note—*
> When multiplying or dividing, the number of significant digits in the result is the same as the least number of significant digits in the calculations.

1.3 Describing Data

You can show the distribution of a set of data graphically using a histogram, a stem-and-leaf plot, or a box-and-whisker plot. You can also describe the distribution numerically using the measures of central tendency: mean, median, and mode. Sometimes, the range and outliers are used to describe the distribution.

Example

According to a national report, students "typically" watch between 3 h and 4 h of TV each day. Paul surveyed his friends, asking the question, "What is the typical amount of time you watch TV each day?" Their responses are shown below. Examine and describe the results of Paul's TV survey.

Amount of Time Students Watch TV Each Day

Name	Hours	Name	Hours
Todd	0 h	Steve	1 h
Richard	3 h	Kinga	4 h
Brian	1 h	Tang	2 h
Craig	2 h	Babatunde	1 h
David	6 h	Joaquin	1 h
Giovanni	9 h	Anson	3 h
Franca	0 h	Albert	4 h
Lily	3 h		

48 Chapter 1 *Data Management*

Solution

Paul's TV Data Represented Graphically

Frequency Table of TV Data

Interval	Frequency
0 to 1	2
1 to 2	4
2 to 3	2
3 to 4	3
4 to 5	2
5 to 6	
6 to 7	1
7 to 8	
8 to 9	
9 to 10	1

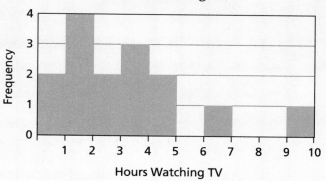

TV Data in a Histogram

Most of the data are between 0 h and 5 h. The 9 h data value seems to be an outlier.

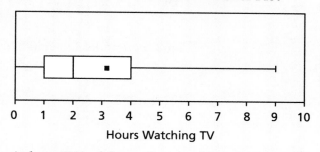

TV Data in a Box-and-Whisker Plot

At least 50% of data values are between 1 h and 4 h.

Paul's TV Data Represented Numerically

Mean

First find the sum of all pieces of data. Then divide by the number of data items.

$$\frac{2 \times 0 + 4 \times 1 + 2 \times 2 + 3 \times 3 + 2 \times 4 + 6 + 9}{15} = 3 \text{ h}$$

The mean time "typically" spent watching TV is 3 h.

Median

To find the median, first put the data values in ascending or descending order.

median
↓
0 0 1 1 1 2 2 3 3 3 4 4 6 9

The "middle" value is the median. (If there is no one middle value, the median is the mean of the two middle values.) The median value is 2 h per week.

Mode
The mode is the value that appears most often.
The value 1 appears four times.
The mode is 1 h per week.

Range
The range gives an indication of the spread of the data.
The range is the difference between the highest and lowest values.
The range is 9 h – 0 h, or 9 h.

Outliers
The value of 9 h may be considered an outlier because it is very different from the other values.

1.4 Defining Data Spread

The **standard deviation** is a measure of the variation of the data. The lower the standard deviation, the less the variation.

The standard deviation is most easily found using technology. If technology is unavailable, a table can be used to help you organize the data.

Example
Find the standard deviation of Paul's set of TV data.

Solution

Recorded Time (h)	Mean	Deviation From the Mean	Squared Deviation
0	3	3	9
0	3	3	9
1	3	2	4
1	3	2	4
1	3	2	4
1	3	2	4
2	3	1	1
2	3	1	1
3	3	0	0
3	3	0	0
3	3	0	0
4	3	1	1
4	3	1	1
6	3	3	9
9	3	6	36

To calculate the standard deviation:
- first, find the sum of the squared deviations.
- then, find the mean of the squared deviations.

- and then find the square root of the mean.

$$\sqrt{\frac{9+9+4+4+4+4+1+1+0+0+0+1+1+9+36}{15}} = 2\text{ h}$$

The standard deviation is 2 h.

1.5 Large Distributions and the Normal Curve

When very large amounts of certain types of data are collected randomly and displayed in a histogram, the frequency polygon appears as a bell-shaped curve that shows a normal distribution.

- It is a symmetric distribution.
- About 68% of the measurements are located within one standard deviation of the mean.
- About 95% of the measurements are within two standard deviations of the mean.
- Outliers are located two standard deviations or more from the mean.
- The mean, median, and mode are the same and are in the middle of the graph.

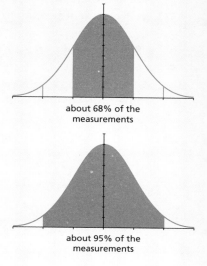

about 68% of the measurements

about 95% of the measurements

1.6 Using Data to Predict

The relationship between two variables can be determined by creating a scatter plot. A line of best fit can be used to make predictions about how the variables relate and predictions about future events.

Example
The table shows the life expectancy for people born in the 20th century.
(a) Make a scatter plot for the data. Draw a line of best fit.
(b) Predict the life expectancy for people born in 1978, 1994, and the year in which you were born.

Birth Year	Life Expectancy (Year)
1900	47.3
1910	50.0
1920	54.1
1930	59.7
1940	62.9
1950	68.2
1960	69.7
1970	70.8
1980	73.7
1990	75.4

Solution

(a) **Life Expectancy by Year of Birth**

(b) 1978 – about 75 years (you are interpolating here)
1994 – about 80 years (you are extrapolating here)

PRACTICE

1.1 Variables and Relationships

1. An experiment is done to study the time it takes for water to evaporate. Thin layers of water are spread on a surface. The surface area is measured and the time it takes for each water pool to evaporate at different temperatures is recorded.

10 cm², 10°C; 20 min
10 cm², 20°C; 10 min
10 cm², 30°C; 5 min
15 cm², 10°C; 5 min
15 cm², 20°C; 2.5 min
15 cm², 30°C; 1.25 min
20 cm², 10°C; 1.25 min
20 cm², 20°C; 0.625 min
20 cm², 30°C; 0.3125 min

   (a) What factors influence the time it takes for a sample of water to evaporate? You can draw a mind map if you wish.
   (b) Arrange the table to show how the time it takes for water to evaporate depends upon the temperature.
   (c) What are the dependent and independent variables?

2. The chart provides the cost for courier delivery according to type of package and distance.
   (a) Organize the data into a table to show how the price of delivering a package depends upon the distance.
   (b) Identify the independent and dependent variables for your table.

Letter, 10 km; $5
Parcel, 50 km; $25
Parcel, 15 km; $14.50
Letter, 5 km; $2.50
Letter, 8 km; $4
Parcel, 100 km; $40
Letter, 7 km; $3.50
Parcel, 70 km; $31

   (c) Describe a model that can be used to predict the delivery charge for various types of packages and letters.
   (d) Predict the cost of delivering a package a distance of 8 km. Does the type of package influence the cost?

1.2 Measuring

3. Answer each problem with the appropriate level of precision.
   (a) A lawn is 12.88 m long and 12.2 m wide. Edging is placed around the lawn. Find the length of edging needed.
   (b) Fred traveled 12.75 km to Anne's house. He stopped for a hamburger on the way home. The restaurant is 5.2 km from Anne's house. How far is Fred from home?
   (c) Find the area of the lawn in (a).

4. Copy these boxes then fill them with numbers that have exactly 4 significant digits:

   ☐.☐☐☐
   ☐.☐☐☐☐
   ☐☐☐ ☐☐☐

5. Find the length, width, and area of this rectangle. Be as accurate and as precise as you can. Report the measurements using an appropriate level of precision.

6. Calculate the area of the shaded region. Record the area using the appropriate level of precision.

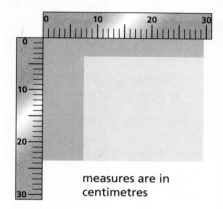

measures are in centimetres

7. How are precision and accuracy alike? How are they different?

1.3 Describing Data

8. A truck is carrying books packaged in boxes. The number of books in each box is recorded as shown.

15	1	20	7	15	8	3	20
8	16	5	4	13	17	20	9
6	16	22	12	6	19	7	9
10	15	9	18	19	15	14	15
2	28	10	17	7	10	8	8

(a) Construct a stem-and-leaf plot, a box-and-whisker plot, a frequency table, and a histogram for the data.

(b) What is the typical number of books in a box? Justify the measure of central tendency you chose.

(c) Describe the spread of the data. How did you make your decision?

(d) Suppose the customer for these books claimed that, "One box had 26 books in it and the box came from your truck. I am not paying for the extra books." Would you believe the customer? Why?

9. (a) Create a set of at least 20 pieces of data that could be described by this box-and-whisker plot.

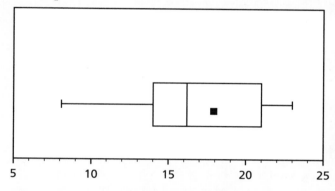

(b) Draw a histogram of your data.

10. Why is it helpful to make a stem-and-leaf plot before you make a box-and-whisker plot? Use an example of your own to help illustrate your answer.

1.4 Defining Data Spread

11. Crates of oranges were packed in two different shipments. The mean diameter in centimetres of the oranges in each crate was recorded in the following tables.

Shipment 1				
9.1	9.0	9.1	9.0	10.0
8.9	8.3	8.8	9.4	8.5
9.2	6.0	8.3	9.2	7.0
9.5	8.5	7.0	9.3	9.0
8.9	9.8	8.7	8.4	9.6
11.0	9.5	9.3	9.7	10.0

Shipment 2				
8.9	9.9	8.4	9.7	6.0
6.1	9.7	7.2	6.8	8.1
6.0	9.3	9.3	8.5	6.8
7.1	8.5	8.9	6.7	8.8
6.4	9.8	7.1	9.6	6.4
8.3	6.2	10.0	7.7	9.1

The shipment of more uniformly sized oranges is to be sold individually but the other shipment is to be made into marmalade. Which shipment will become marmalade? Justify your decision.

12. Leif conducted some tests to determine whether regular gas or premium gas resulted in better mileage. He kept track of how many kilometres he was able to drive on each full tank of gas. He repeated the tests many times using the two different types of gas and recorded the following results. Is there a significant difference?

Regular Gas		
640	570	660
580	610	540
555	588	615
570	550	590
585	587	591

Premium Gas		
659	619	639
629	664	635
709	637	633
618	604	638
689	589	599

13. Sketch two histograms, one in which the standard deviation would be small (but greater than 0) and one in which it would be quite large.

1.5 Large Distributions and the Normal Curve

14. How can you use the mean and standard deviation of normally distributed measurements to decide if a measurement is an outlier?

15. Why might you need to graph a lot of measurements to get a histogram that looks like a normal curve?

1.6 Using Data to Predict

16. The scatter plot shows how the distance traveled changed with the amount of gas used.
    (a) Copy the graph. Construct a line of best fit.
    (b) Use the line of best fit to predict the distance driven on 70 L of fuel.

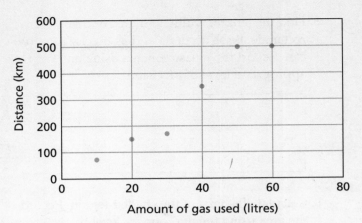

   (c) Use the line of best fit to predict the distance driven on 25 L of fuel.
   (d) Identify the processes used in parts (b) and (c) as interpolation or extrapolation. Give reasons for your choice.
   (e) Suggest reasons why the data points in the graph do not lie exactly in a straight line.

17. Mary's mother has kept a record of Mary's height on her birthday since she was born.

Age (years)	Height (cm)
0	50
1	64
2	75
3	90
4	105
5	120
6	128
7	135

   (a) Draw a scatter plot to show Mary's height on each birthday.
   (b) Construct a line of best fit. What do you think Mary's height was when she was 4.5 years old?
   (c) Would your line be a good predictor of Mary's height when she turns 18? Explain. Are there any limitations on using your line of best fit?

54 Chapter 1 *Data Management*

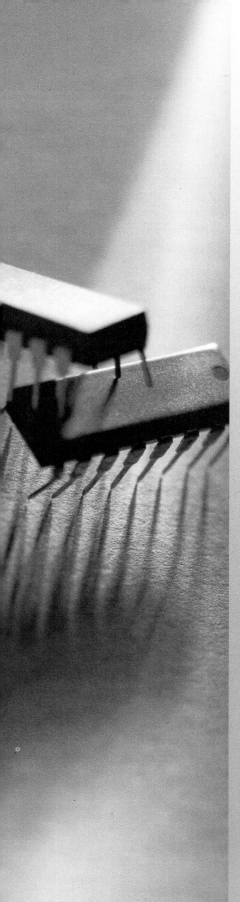

Chapter Two
Networks and Matrices

Information can be represented in various ways, and some ways are easier to interpret than others. For example, sets of collected data are often represented graphically. This enables you to see how the data are distributed, or to see relationships between variables.

In this chapter, you will learn about ways to represent information about networks using diagrams called graphs. Information about networks of flights between airports is often represented in a "map" or graph. Tables or matrices will be used to represent information about networks as well as other types of information. Once information is in graph or matrix form, it is easier to solve related problems.

After successfully completing this chapter, you will be expected to:

1. Represent networks with appropriate graphs.
2. Interpret network graphs to solve related problems.
3. Represent digraphs with adjacency matrices and vice versa.
4. Multiply matrices to solve related problems.
5. Multiply adjacency matrices to solve network problems.

2.1 Creating and Traveling Network Graphs

network – a set of people, places, objects, or ideas that are connected in some way

If you have a network of friends, it means that you are "connected" to all the individuals in that group by friendship. A network is a set of people, places, objects, or ideas that are connected in some way. Other examples of networks are transportation networks, computer networks, and networks of streets. In this section, you will use diagrams, or graphs, to represent networks and to solve related problems.

Think about...

Graphs

Where else have you created a graph in order to solve a problem?

The A-Mazing Challenge

Leigh is using her new computer game, A-Mazing. She must travel every path of the maze exactly once. If she tries to go over a path a second time, the game ends and she loses. She is shown a map of the maze in advance so that she can plan. Once she begins, she has only 30 s to get through the whole maze, so she has to plan carefully. Is it possible to get through the maze without going over any path twice?

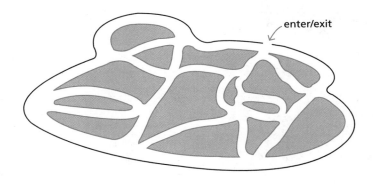

Think about...

The Maze

1. Does it matter if the paths are straight or wiggly in deciding whether the maze can be traveled as required?

2. Suppose there had been one extra path included in the maze, as shown in the diagram below. Would Leigh still be able to travel the maze as required? How?

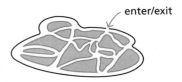

3. The following diagram also represents the maze. Why?

Investigation 1

Can You Cover a Network Completely, Traveling Each Edge Exactly Once?

Purpose

In this Investigation, you will use network graphs to solve mail-delivery problems.

Mail carriers try to cover a network of roads traveling the minimum distance. That means finding a route that does not repeat any roads, if possible.

Procedure

A. Each **network graph** represents roads for mail delivery. Which networks can be traveled without repeating a road? Record your route. (Note: Assume that mail boxes are on one side of each road.)

(i)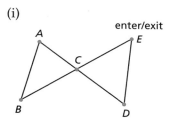

(ii)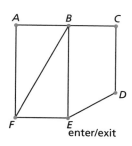

(iii)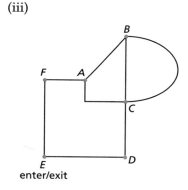

(iv)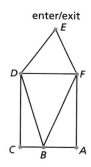

(v)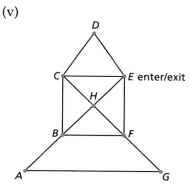

Why do you think that some of the networks can be traveled without repeating a road while others cannot?

network graph – consists of a set of points called **vertices** (singular is **vertex**) connected by lines called **edges**

Think about...

Step A

Why are all vertices identified by letters?

To speed up delivery in rural neighbourhoods, why should all mail boxes be located on one side of the road?

—Note—
There are no arrows on the edges. This indicates that the roads can be traveled in either direction. A mail carrier will choose the direction he or she wishes to travel on a road.

B. For each network in A, how many roads meet at each vertex? Copy each network graph in your notes and record your answers at each vertex of your diagram.

C. Examine the results for the networks that can be traveled without repeating a road. What do you notice?

D. Create two more networks that can be traveled without repetition.

Investigation Question

1. (a) Examine the networks that cannot be traveled without repeating a road. Add one edge to each graph so that a non-repeating route is possible. What do you notice?

   (b) In each of the network graphs in Step A, you entered and exited the network from the same vertex. For some of these networks, it was possible to do this while traveling each edge only once. How can you quickly determine if a network is Eulerian without tracing a route?

   (c) Test your conjecture in (b) using several network graphs.

   (d) Look again at the networks that cannot be traveled completely without repeating a road. Can any of these be traveled completely without repeating if you enter and exit at different vertices?

Eulerian – a network is Eulerian if you can enter and exit it from a single vertex while traveling each edge only once

Check Your Understanding

— Note —
All streets are edges and intersections are vertices.

2. These network graphs represent four neighbourhoods to which Trevor delivers newspapers. He rides a bike along each street and tosses the newspapers onto the driveways of houses on both sides of the street. Which neighbourhood(s) can Trevor travel without repeating a street if:

   (a) he enters and exits at the same intersection?

   (b) he enters and exits at different intersections?

   A   B

even vertex – a vertex at which an even number of edges meet

odd vertex – a vertex at which an odd number of edges meet

C   D

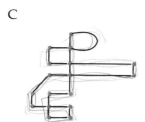 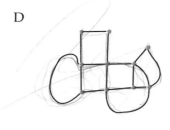 

3. A security guard at a campground patrols the park on foot. She must cover all the roads, but does not want to walk any farther than she has to. Can she travel the network without repeating a road? Explain.

Campground Security Patrol Network

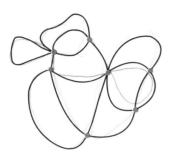

4. Create three network graphs as follows:
   (a) one that can be traveled completely without repeating an edge, entering and exiting at the same vertex;
   (b) one that can be traveled completely without repeating an edge, entering and exiting at different vertices; and
   (c) one that cannot be traveled completely without repeating an edge.

5. A parking-meter officer must check all parking meters in a busy downtown area. The meters are shown as dots on the map.

Parking Meter Network

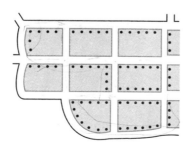 

> **Did You Know?**
>
> Leonhard Euler (1707–1783) was a famous Swiss mathematician who studied networks and other related problems. He is famous for the Seven Bridges of Königsberg. Euler wondered if one could travel across all seven bridges exactly once, beginning and ending at the same point on the mainland. As a result, such a network is Eulerian.
>
>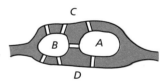
>
> Bridges of Königsberg
>
> You will discover more about Leonhard Euler and the Seven Bridges of Königsberg in the Practice, Section 2.1, question 4.

2.1 Creating and Traveling Network Graphs   59

(a) Draw a graph to represent the network of paths connecting the meters.

(b) Can the officer check all the meters without repeating any streets, entering and exiting at the same intersection? Explain.

(c) Can the officer check all the meters without repeating any streets, by entering and exiting at different intersections? If so, explain.

6. Return to the network graphs in Investigation 1, Step A.
   (a) Copy Graphs (i) and (ii).
   (b) Suppose the mailboxes were located on both sides of each road so that each road must be traveled twice. Revise each graph to represent the new network.
   (c) Which of the new networks can be traveled without repeating an edge, entering and exiting at the same vertex?
   (d) What do you notice?

7. A garbage truck can collect garbage in one trip down each street.
   (a) Describe its route for each network.
   (b) For which networks can the garbage truck begin and end at the same intersection?
   (c) Suppose the garbage truck needed to make two trips down each street. For which networks can the garbage truck begin and end at the same intersection?

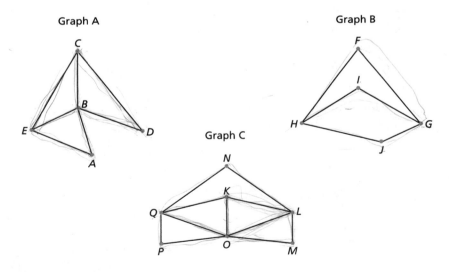

8. A cross-country ski club sets one track on each trail and is shown at the top of the next page.
   (a) Describe the best route for setting the tracks, beginning and ending at the warm-up cabin. Will any trails have to be traveled more than once?

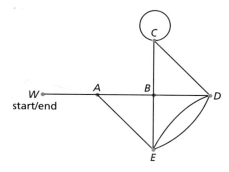

(b) Create a network of trails. Design it so that the tracks can be set by traveling each trail only once, beginning and ending at the same vertex. Describe the route.

(c) For your network, describe a route for setting two sets of tracks on each trail.

9. A desirable feature in a house design is easy access through various parts of the house.
   (a) Draw an access graph for each house plan.
   (b) Can you tour the entire house without going through a doorway twice? Explain.

House Plan 1

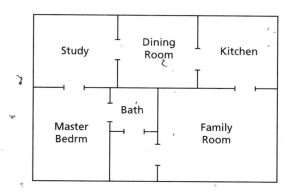

House Plan 2

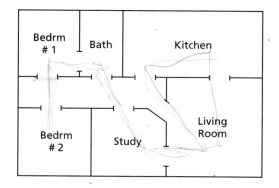

Did You Know?

Sir William Rowan Hamilton (1805–1865) created a graph game that required players to visit 20 major cities in Europe without visiting any city twice. His game, called the Icosian game, was played on a regular dodecahedron, a 3-D shape with 20 vertices.

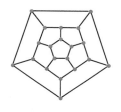

— Note —
Architects often draw diagrams to provide a quick view of routes between rooms of a house. The diagrams that represent the network of possible routes are called access graphs. In these diagrams, rooms are vertices and doorways are edges.

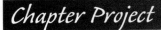

Designing a Park

Your project is to plan a park that includes items such as:

- bicycle and/or hiking trails;
- playground equipment;
- gardens and ponds; and
- a water slide or amusement rides.

You park design must be carefully planned. Your final presentation should demonstrate an understanding of the mathematical content from this chapter.

Here are some questions to get you started:

- Where will you locate your park? It may be located in an urban or rural setting.
- How much area will it cover? It can be as small as 0.5 ha or as large as 100 ha.
- Does it have any physical attributes that might affect the design? For example, does it have a small brook or ravine?
- How will you represent your park? For example, will you use a scale drawing, a computer program, or a 3-D model?

Section 2.1 Connections

(a) Explain why bicycle or hiking trails can be represented by network graphs.

(b) What would the vertices and edges of a trail network represent?

(c) For a cyclist in the park, which would be of more interest: traveling edges or visiting vertices? Explain.

(d) Explain why a network of playground equipment can be represented by a network graph.

(e) What would the vertices and edges of a playground network represent?

(f) For a child playing at the playground, which would be of more interest: traveling edges or visiting vertices? Explain.

2.2 Digraphs and Adjacency Matrices

Another computer maze game, Mazeplus, requires the player to travel each path through the maze only once. This time, each path is one way and must be followed in the direction indicated.

Look at each intersection in the maze shown below. If two paths lead toward an intersection in the maze, how many paths must lead away from that location if no paths can be repeated? Why? Is it possible to travel this maze without repeating a path?

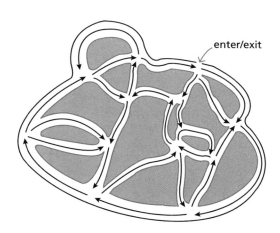
enter/exit

Think about...

The Maze

Can you change the Mazeplus maze so that it can be traveled without repeating a path?

Investigation 2
Can All Edges Be Traveled?

Purpose

In this Investigation, you will explore situations in which you can travel only in specific directions. These can be represented by directed graphs, or **digraphs**. Arrows are used along the edges to indicate the allowed direction.

The following graphs represent mazes where each path must be traveled, only once, and in the direction indicated. The exit location may or may not be the same as the entrance.

digraph – a directed graph. The edges of the graph can be traveled only in the indicated direction.

Procedure

A. For each maze, decide whether all paths can be traveled without repeating a path.

(i)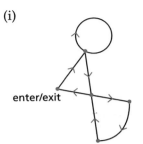

(ii)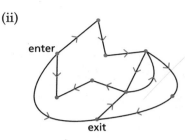

(iii)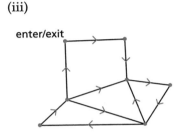

(iv)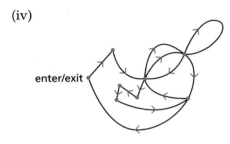

(v)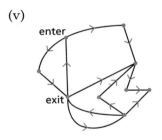

(vi)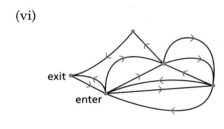

B. If the maze cannot be traveled without repeating a path, explain where the problem is and why. Draw extra paths to solve the problem.

Investigation Questions

1. Look at the digraphs that have the entrance and exit at the same vertex. Name the vertices on these digraphs. At each vertex, record the number of paths leading in, the number leading out, and the total number. What do you notice?

2. Look at the digraphs that have the entrance and exit at different vertices. Name the vertices as in Question 1. At each vertex, record the number of paths leading in, the number leading out, and the total number. What do you notice?

Questions 1 and 2

What conditions are needed for a maze to be traveled without repeating any paths?

Check Your Understanding

3. Each network diagram shows the passes that led to a goal during a hockey game.

Goal 1

Goal 2

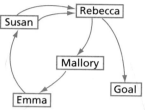

(a) Re-draw each diagram as a digraph.
(b) What do the vertices and paths represent?
(c) For each goal, state who passed first and who scored.

 ## Simplifying Digraphs

A graph becomes complicated when there are many directed paths. There are ways to simplify the graph and display the same information.

Example
This digraph represents the network of direct flights among three airports: Charlottetown (C), Deer Lake (D), and Edmunston (E). Simplify it as much as possible.

Direct Flight Digraph

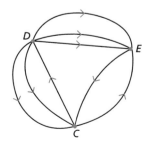

Solution

The digraph can be simplified as follows:

- The three paths *from* D *to* E have been simplified to one path *from* D *to* E labeled "3."
- The two paths *from* D *to* C are now one path *from* D *to* C labeled "2."

Direct Flight Digraph

Original   Simplified

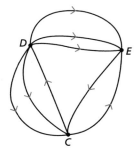 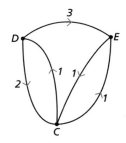

> — Note —
> Digraphs with many directed edges between adjacent vertices can be simplified. You can use numbers to represent the number of edges going in the same direction between two vertices.

Because there are the same number of paths going in both directions between E and C, the graph can be simplified even more:

Direct Flight Digraph—Simplified

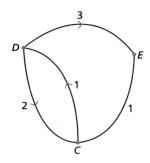

> — Note —
> There is a number 1 but no direction indicated on the edge between E and C. This means that there is the same number of flights going both ways; in this case, one flight each way.

Focus Question

4. (a) Draw a graph to show two flights from E to D.
   (b) Suppose three flights from E to D were added to the original. How would this change the simplified graph?

66 Chapter 2 *Networks and Matrices*

Check Your Understanding

5. Simplify the following digraphs.

(a)

(b)

(c)

6. (a) Create a digraph that includes:
  - a path that is one way;
  - a path that is one way and marked with the number 2;
  - a path that is two way; and
  - a path that is two way and marked with the number 2.

(b) Then redraw the graph so that it shows the same information, but no numbers appear on any paths.

Representing Digraphs as Adjacency Matrices

From Digraph to Table

The information in a digraph can also be represented in table, or **matrix**, form. For example, recall the simplified flight digraph from Focus A.

matrix – a rectangular array of numbers (plural is matrices)

Direct Flight Digraph

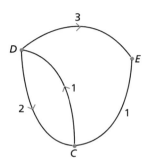

— Note —
When interpreting a matrix, you must still keep the headings in mind.

To represent the same information in a table:

Step 1 Create the same headings for both rows and columns using the labels for the vertices.

Step 2 Complete each row by entering the direct flights from each airport in each row to each airport in each column.

	TO		
FROM	C	D	E
C	0	1	1
D	2	0	3
E	1	0	0

From Table to Adjacency Matrix

A matrix is a table without row and column headings. The matrix form for the direct flight table is:

$$\begin{pmatrix} 0 & 1 & 1 \\ 2 & 0 & 3 \\ 1 & 0 & 0 \end{pmatrix}$$

The element 2 is located in the second row, first column.

Each number in a matrix is an **element**. The above matrix has nine elements. The **dimensions** of this matrix are 3 rows by 3 columns, or in short, 3 by 3.

The matrix of the digraph is a special type of matrix called an **adjacency matrix**, because it shows connections between adjacent vertices in a network.

Focus Questions

7. (a) What is the total number of direct flights between Deer Lake and Charlottetown?
  (b) How can you tell from looking at the digraph?
  (c) How can you tell from looking at the adjacency matrix?

8. (a) Which community has the greatest number of departures? arrivals?
  (b) How can you tell from looking at the digraph?
  (c) How can you tell from looking at the adjacency matrix?

Think about...

Rows 2 and 3
What does each number in rows 2 and 3 tell you about direct flights between airports?

element – a number in a matrix, for example, the matrix on this page has nine elements

dimensions – the size of a matrix, described as the number of rows by the number of columns

adjacency matrix – a special matrix that organizes information about connections between adjacent vertices in a network

9. (a) It is possible to fly from Edmunston to Deer Lake? Describe how.
   (b) If it is possible, why is the element in row 3, column 2 a zero?

10. Explain why there are zeros along the diagonal of the adjacency matrix.

11. Suppose the airline added a sight-seeing tour departing from and returning to Edmunston.
    (a) Draw the new digraph.
    (b) Create the new matrix.
    (c) Compare the original and revised matrices. What do you notice?

Think about...

Adjacency Matrices
Why are adjacency matrices always square?

Check Your Understanding

12. This digraph represents a network of cross-country ski trails.

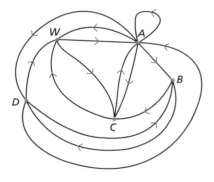

    (a) Describe the possible routes for skiing from the Warming Cabin (W) to the Cedar Cabin (C).
    (b) How many direct routes are there from W to C?
    (c) Create a matrix to show the number of direct routes between vertices.
    (d) What are the dimensions of the matrix?
    (e) What element of the matrix shows the number of direct routes from W to C?
    (f) What does the element in row 4, column 2 represent?

13. (a) Simplify this digraph.

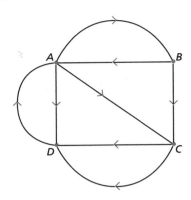

(b) Create an adjacency matrix to represent the digraph.

14. (a) Copy and complete the table on page 71 that corresponds to this flight digraph.

Direct Flights Within Newfoundland and Labrador Each Day

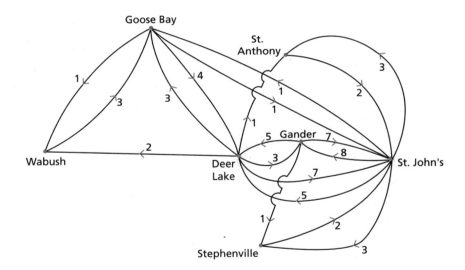

Question 14

Where do you think this airline should locate its headquarters? Explain.

	TO						
FROM		Deer Lake	Gander	Goose Bay	St. Anthony	St. John's	Stephenville	Wabush
	Deer Lake							
	Gander							
	Goose Bay							
	St. Anthony							
	St. John's							
	Stephenville							
	Wabush							

(b) Present the information on this table as an adjacency matrix.

(c) What are the dimensions of the matrix?

(d) What is the element in row 7, column 3? What does it represent?

(e) Explain why all the elements along the **major diagonal** of this matrix are zero.

major diagonal — the diagonal of a square matrix that goes from the top left to the bottom right

(f) Which communities are not connected by direct flights? How can you tell from looking at the digraph? The adjacency matrix?

(g) Which community has the greatest number of departures daily? How many departures are there? Does it also have the greatest number of arrivals?

15. (a) Create a digraph for this direct flight matrix.
    (b) Simplify the digraph as much as possible.
    (c) How did you know from the matrix that there would be a loop in the digraph?
    (d) How could you check your digraph to make sure nothing is missing?

$$\begin{pmatrix} 0 & 4 & 2 \\ 4 & 0 & 1 \\ 3 & 1 & 1 \end{pmatrix}$$

16. Create the adjacency matrix for each digraph.

    Digraph A

    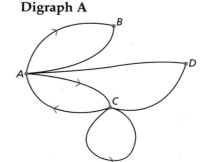

    Digraph B

    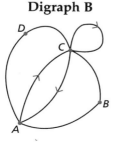

— Note —
For Question 16, when an edge has no number indicated, assume it is labeled with the number 1.

Digraph C

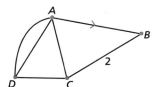

Digraph D

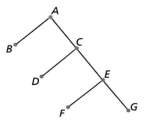

17. Draw a digraph for each adjacency matrix.

(a) $\begin{pmatrix} 1 & 0 & 0 \\ 0 & 1 & 0 \\ 0 & 0 & 1 \end{pmatrix}$

(b) $\begin{pmatrix} 2 & 2 & 0 \\ 0 & 0 & 2 \\ 2 & 2 & 1 \end{pmatrix}$

(c) $\begin{pmatrix} 1 & 1 & 1 & 1 \\ 1 & 1 & 1 & 1 \\ 1 & 1 & 1 & 1 \\ 1 & 1 & 1 & 1 \end{pmatrix}$

(d) $\begin{pmatrix} 0 & 1 & 0 & 0 & 0 & 1 \\ 1 & 0 & 1 & 0 & 0 & 0 \\ 0 & 1 & 0 & 1 & 0 & 0 \\ 0 & 0 & 1 & 0 & 1 & 0 \\ 0 & 0 & 0 & 1 & 0 & 1 \\ 1 & 0 & 0 & 0 & 1 & 0 \end{pmatrix}$

(e) $\begin{pmatrix} 0 & 1 & 0 \\ 1 & 1 & 1 \\ 0 & 1 & 0 \end{pmatrix}$

18. Electrical systems are connected in large networks. During emergencies or times of high demand, one company may sell power to another company in the network. Instead of flowing directly from the source to the receiving station, the power divides, following several routes in the network. This adjacency matrix describes the direct routes possible. Draw a digraph to represent the matrix.

$$\text{FROM} \begin{array}{c} \\ A \\ B \\ C \\ D \end{array} \overset{\text{TO}}{\begin{pmatrix} A & B & C & D \\ 0 & 1 & 1 & 1 \\ 0 & 0 & 0 & 1 \\ 1 & 0 & 0 & 1 \\ 1 & 1 & 1 & 0 \end{pmatrix}}$$

19. Regular bus shuttle service is available as shown in the digraph. Construct an adjacency matrix to represent this digraph.

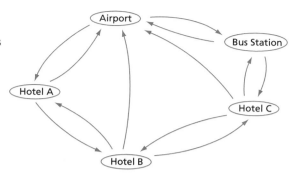

20. Part of a forest ecosystem is described below:
    - plants are eaten by insects, caterpillars, and songbirds;
    - insects are eaten by toads, songbirds, spiders, hawks, and one another;
    - spiders are eaten by songbirds and other spiders;
    - toads and songbirds are eaten by hawks; and
    - caterpillars are eaten by songbirds.
    (a) Draw a digraph to represent this food web.
    (b) Represent this food web using an adjacency matrix.

21. In this Section and in Section 2.1, you have been introduced to many different types of situations that can be represented by graphs and adjacency matrices.
    (a) Review the two sections and create a list of those situations. Add any other situations you can think of.
    (b) Select several networks from your list and state what the vertices and paths would represent.

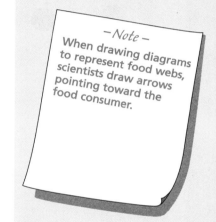

— Note —
When drawing diagrams to represent food webs, scientists draw arrows pointing toward the food consumer.

Chapter Project

Designing a Park

Continue to work on your park design. Use these questions to demonstrate how you can apply what you have learned in this section.

Section 2.2 Connections

(a) Represent this adjacency matrix as a digraph of bike trails.

$$\begin{array}{c} \\ A \\ B \\ C \end{array} \begin{pmatrix} A & B & C \\ 1 & 1 & 0 \\ 0 & 0 & 1 \\ 1 & 1 & 0 \end{pmatrix}$$

(b) Why might there be directed paths on a digraph of a network of bike trails?

2.3 Matrix Multiplication

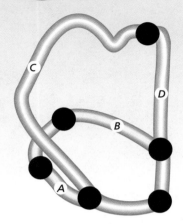

The computer game Touring requires two players to compete. Each player starts at a random location decided by the computer, and must travel along the edges shown. Points are awarded if edges A, B, C, or D are traveled, according to their lengths as shown in the chart below. Sometimes an obstacle comes up and the player is blocked from using the path he or she wants. The player with the most points wins.

Investigation 3
A Game of Touring

Purpose

In this Investigation, you will explore how to find the number of points scored by players in a game of Touring.

Procedure

Here are the results for Manny and Nicole when they played Touring.

Game 1

Number of Times
Path Traveled

Path	A	B	C	D
Manny	2	2	2	0
Nicole	0	0	3	2

Points Earned
for a Path

Path	Points
A	2
B	2
C	4
D	3

A. Find the number of points each player received. Who won the game?

The Result of the Game

Why is it not surprising who won?

Chapter 2 *Networks and Matrices*

Investigation Questions

1. To find Nicole's score, what numbers did you multiply? What numbers did you add? What do they represent? How do you know?

2. To find Manny's score, what numbers did you multiply? What numbers did you add? What do they represent? How do you know?

3. Record the tables from Game 1 as well as your answer for the total scores in matrix form. The column titles for the matrices are shown. Label each row.

Number of Times 　　Points Earned 　　Score Earned by
Path Traveled 　　　for a Path 　　　　Each Player

$$\begin{pmatrix} A & B & C & D \\ 2 & 2 & \_ & \_ \\ 0 & \_ & \_ & \_ \end{pmatrix} \quad \begin{pmatrix} 2 \\ \_ \\ \_ \\ \_ \end{pmatrix} \quad \begin{pmatrix} \_ \\ \_ \end{pmatrix}$$

4. Suppose that Manny and Nicole traveled the paths in the same way as before, but the points awarded to each path were changed to the following:

Game 2

Path	Points
A	2
B	1
C	4
D	2

(a) Use this information to complete three matrices as shown below. Label each row.

Number of Times 　　Points Earned 　　Score Earned
Path Traveled 　　　for a Path 　　　　by Each Player

$$\begin{pmatrix} \_ & \_ & \_ & \_ \\ \_ & \_ & \_ & \_ \end{pmatrix} \quad \begin{pmatrix} \_ \\ \_ \\ \_ \\ \_ \end{pmatrix} \quad \begin{pmatrix} \_ \\ \_ \end{pmatrix}$$

(b) Who won in this situation?

5. If the "points earned for an edge" matrix for Games 1 and 2 are combined into one matrix, it would look like this:

$$\begin{pmatrix} 2 & 2 \\ 2 & 1 \\ 4 & 4 \\ 3 & 2 \end{pmatrix}$$

What heading would go at the top of each column?

6. (a) The information for Games 1 and 2 is given in the matrices below. Complete the third matrix.

Number of Times Path Traveled

$$\begin{array}{c} \\ \text{Manny} \\ \text{Nicole} \end{array} \begin{pmatrix} A & B & C & D \\ 2 & 2 & 2 & 0 \\ 0 & 0 & 3 & 2 \end{pmatrix}$$

Points Earned for a Path

$$\begin{array}{c} \\ A \\ B \\ C \\ D \end{array} \begin{pmatrix} G.1 & G.2 \\ 2 & 2 \\ 2 & 1 \\ 4 & 4 \\ 3 & 2 \end{pmatrix}$$

Score Earned by Each Player

$$\begin{array}{c} \\ \text{Manny} \\ \text{Nicole} \end{array} \begin{pmatrix} \square & \square \\ \_\_ & \_\_ \\ \_\_ & \_\_ \end{pmatrix}$$

(b) What might the labels be at the top of each column in the third matrix?

(c) There is a 14 in the third matrix. What does it represent? Describe its location in the matrix using row and column numbers.

(d) What row and what column of numbers were multiplied and added to obtain the 14? To obtain the 18?

(e) How do you know how to place the numbers in the third matrix? Describe your pattern.

(f) Record the dimensions of the three matrices.

Think about...

Question 6

Why does there have to be the same number of rows in the second matrix, telling the points earned for a path, as the number of columns for the first matrix telling the number of times each path was traveled?

Check Your Understanding

7. Pierre and Helen played a different Touring game. The matrices below describe their game.

Number of Times Path traveled

$$\begin{array}{c} \\ \text{Pierre} \\ \text{Helen} \end{array} \begin{pmatrix} X & Y & Z \\ 3 & 2 & 1 \\ 0 & 3 & 2 \end{pmatrix}$$

Points Earned for a Path

$$\begin{array}{c} \\ X \\ Y \\ Z \end{array} \begin{pmatrix} \text{Game} & 1 & 2 & 3 \\ & 1 & 2 & 3 \\ & 2 & 4 & 6 \\ & 3 & 1 & 5 \end{pmatrix}$$

Score Earned by Each Player

$$\begin{pmatrix} \square & \square & \square \\ 10 & 15 & 26 \\ 12 & 14 & 28 \end{pmatrix}$$

(a) Copy and label the third matrix.

(b) How many paths earned points in this game?

(c) How many different point systems were used?

(d) Explain what the 26 represents and how it was obtained.

(e) Which element is at the row 2, column 1 location in the third matrix? What does it represent?

(f) What are the dimensions of these three matrices?

8. Richard played two other Touring games by himself. The information on the "number of times each path was traveled" and "the points earned for each path" are shown in the matrices below.

Path   M N O P Q

Game 1 $\begin{pmatrix} 3 & 0 & 4 & 2 & 5 \\ 2 & 1 & 0 & 6 & 3 \end{pmatrix}$
Game 2

$\begin{pmatrix} \square & \square \\ \square & 1 & 2 \\ \square & 1 & 2 \\ \square & 2 & 1 \\ \square & 2 & 1 \\ \square & 1 & 3 \end{pmatrix}$

(a) Copy and label the second matrix.

(b) How many paths earned points in this game?

(c) How many games were played?

(d) Determine the third matrix showing Richard's score for each game. Label the matrix.

(e) What are the dimensions of each of the three matrices?

(f) What pattern do you notice about the matrix dimensions for Questions 6, 7, and 8?

Product Matrices

— Note —
This relationship between the dimensions of the matrices being multiplied and the dimensions of the product matrix can be expressed using algebra.
(*m* by *n*) × (*n* by *p*)
= (*m* by *p*)

To find the element in a certain row and a certain column, you pair up numbers in that row of the first matrix and that column of the second matrix, multiply the pairs, and then add the products.

For example, a 2 by 4 matrix multiplied by a 4 by 1 matrix results in a 2 by 1 product matrix.

$$\begin{pmatrix} 2 & 2 & 2 & 0 \\ 0 & 0 & 3 & 2 \end{pmatrix} \times \begin{pmatrix} 2 \\ 2 \\ 4 \\ 3 \end{pmatrix} = \begin{pmatrix} 2\times 2+2\times 2+2\times 4+0\times 3 \\ 0\times 2+0\times 2+3\times 4+2\times 3 \end{pmatrix} = \begin{pmatrix} 16 \\ 18 \end{pmatrix}$$

2 by 4   × 4 by 1   = 2 by 1

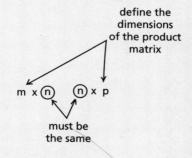

define the dimensions of the product matrix

m x (*n*)   (*n*) x *p*

must be the same

The product matrix has the row dimensions of the first matrix and the column dimensions of the second matrix.

Focus Questions

> **— Note —**
> Remember: the dimensions of a matrix are the number of rows by the number of columns.

9. Multiply the following pairs of matrices on your calculator. Describe how your product matrix is connected to the pair of original matrices.

   Pair A
   $$\begin{pmatrix} 1 & 2 \\ 4 & 5 \\ 7 & 8 \end{pmatrix} \qquad \begin{pmatrix} 2 & 3 \\ 4 & 5 \end{pmatrix}$$

   Pair B
   $$\begin{pmatrix} 1 & 2 & 3 \\ 4 & 5 & 6 \\ 7 & 8 & 9 \\ 10 & 11 & 12 \end{pmatrix} \qquad \begin{pmatrix} 2 & 3 \\ 5 & 6 \\ 4 & 1 \end{pmatrix}$$

10. Explain, in writing, how to tell if two matrices can be multiplied.

11. Describe a general procedure for multiplying two matrices.

Check Your Understanding

12. The matrix below shows the number of air miles you would collect if you flew from Fredericton to three different locations.

    Air Miles

    Fredericton to Toronto $\begin{pmatrix} 650 \\ 500 \\ 500 \end{pmatrix}$
    Fredericton to Montreal
    Fredericton to Halifax

    A person made a number of flights, some in the winter and some in the summer, as shown in this matrix.

    Number of Flights

	Fredericton to Toronto	Fredericton to Montreal	Fredericton to Halifax
Winter	2	4	6
Summer	8	6	4

Chapter 2 *Networks and Matrices*

(a) These matrices can be multiplied. Explain how.

(b) Multiply the matrices.

(c) How many air miles are earned in the winter and in the summer?

13. Paul has a bike-repair shop in his basement. He has a record of inventory by component.

$$\begin{array}{c} \text{Tires} \quad \text{Handlebars} \quad \text{Pedals} \quad \text{Frames} \\ \text{Number of Items} \quad \begin{pmatrix} 5 & 3 & 6 & 2 \end{pmatrix} \end{array}$$

Paul has estimated an average value for each component and has organized the data in another matrix, shown below.

$$\begin{array}{c} \text{Average Value (\$)} \\ \begin{array}{l} \text{Tires} \\ \text{Handlebars} \\ \text{Pedals (each)} \\ \text{Frames} \end{array} \begin{pmatrix} 20 \\ 30 \\ 25 \\ 300 \end{pmatrix} \end{array}$$

(a) Paul has decided to triple his present inventory of bike parts for next season. What will the new inventory matrix look like?

(b) Find the total value of the inventory for next season.

(c) How many elements are in the product matrix? Is this surprising? Why or why not?

14. Two of these matrices have been multiplied to create the third matrix. Identify the first matrix, the second matrix, and the product matrix.

Matrix A
$$\begin{pmatrix} 123 & 37 & 108 \\ 36 & 8 & 30 \end{pmatrix}$$

Matrix B
$$\begin{pmatrix} 2 & 4 & 7 \\ 5 & 1 & 3 \\ 8 & 0 & 4 \end{pmatrix}$$

Matrix C
$$\begin{pmatrix} 7 & 9 & 8 \\ 2 & 0 & 4 \end{pmatrix}$$

15. Two **square matrices** can be multiplied together. What do you know:

(a) about the dimensions of the two matrices? Explain how you know.

(b) about the dimensions of the product matrix? Explain how you know.

— Note —
When you multiply each element in a matrix by the same number, as in Question 13(a), it is called multiplying a matrix by a scalar.

square matrix – a matrix in which the number of rows equals the number of columns

For example:

$$\begin{pmatrix} 2 & 3 \\ 1 & 2 \end{pmatrix} \text{ and } \begin{pmatrix} 4 & 3 & 5 \\ 1 & 2 & 0 \\ 0 & 3 & 1 \end{pmatrix}$$

are square matrices.

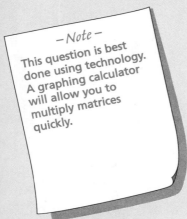

— Note —
This question is best done using technology. A graphing calculator will allow you to multiply matrices quickly.

16. At a photo store's two outlets, the inventory of camera models A, B, and C is to be sold in two ways:
    - at the regular price; and
    - at a promotional price to include three films and processing.

    Number of Cameras

Store	Camera Model		
	A	B	C
Ryder Rd	40	30	25
Beck Street	35	20	15

	Regular $	Promo $
A	$90.35	$102.80
B	$193.25	$201.90
C	$292.60	$302.65

    (a) Use matrix multiplication to show the total value of the cameras at the Beck Street outlet if they are sold at the regular price.
    (b) Use matrix multiplication to show the total value of all cameras at both outlets if they are sold at the regular price.
    (c) Use matrix multiplication to show the total value of all cameras at both outlets if they are sold at the promotional price.

17. (a) Create two matrices that can be multiplied. Label the rows and columns in each matrix to show what the elements might represent.
    (b) Find the product matrix and describe its information.

Investigation 4
Using Matrix Multiplication to Solve Network Problems

This adjacency matrix shows direct flights between three airports.

Matrix A
Direct Flight Matrix

$$\begin{array}{c} \\ \text{Deer Lake} \\ \text{Edmunston} \\ \text{Charlottetown} \end{array} \begin{array}{ccc} \text{Deer Lake} & \text{Edmunston} & \text{Charlottetown} \end{array} \\ \begin{pmatrix} 0 & 1 & 1 \\ 1 & 0 & 1 \\ 1 & 1 & 0 \end{pmatrix}$$

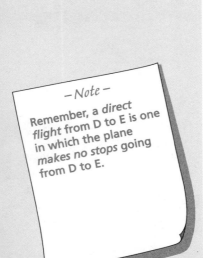

— Note —
Remember, a direct flight from D to E is one in which the plane makes no stops going from D to E.

Purpose
In this Investigation, you will use matrix multiplication to determine the number of flights between airports with exactly one stopover.

Procedure
A. Draw a digraph based on the adjacency matrix.

B. Suppose you started in Deer Lake, where could you go next? Then where could you go? Create a matrix that shows all flights between locations that have exactly one stopover.

C. Multiply the *Direct Flight Matrix* (Matrix A) by itself, that is, square Matrix A. Label the product matrix *Matrix A²*.

D. Compare your *One-Stopover Flight Matrix* from Step B and *Matrix A²* from Step C. What do you notice?

– Note –

In part B, for example, in the Deer Lake row, in the Deer Lake column, include the flight from Deer Lake back to Deer Lake that stops once in Edmunston. Label your new matrix One-Stop Flight Matrix.

Investigation Question

18. (a) Predict what sort of information would be included in Matrix A^3.
    (b) Check your prediction by using the digraph and by cubing Matrix A.

Check Your Understanding

19. Examine the direct-flight digraph to the right.

    (a) Use the digraph to determine the number of possible one-stopover flights from C to D with a stop in E. Explain how you arrived at your answer.

    (b) How many possible flights are there between C and B with a stop at A? Explain how you know.

    (c) How many one-stop flights are there from A to D? Explain how you arrived at your answer.

    (d) Construct an adjacency matrix for the digraph. Then square it.

    (e) Use the matrix to determine the number of one-stop flights from A to D.

20. (a) Create a digraph with four vertices.
    (b) Create the adjacency matrix for your digraph.
    (c) Square your adjacency matrix.
    (d) Explain how each element along the diagonal of the squared matrix relates to your digraph.

Direct Flights Among Five Airports

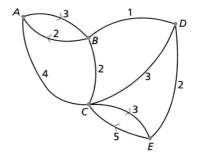

Think about...

Step C
Why is it possible to square an adjacency matrix? Create a matrix that you cannot square. Why can it not be squared?

21. Matrix A indicates whether you can fly between certain Atlantic locations.

Matrix A

Direct Flights Between Airports in Atlantic Canada

$$\begin{array}{c} \\ \text{Fredericton} \\ \text{Moncton} \\ \text{Charlottetown} \\ \text{Halifax} \\ \text{Sydney} \\ \text{Deer Lake} \\ \text{St. John's} \end{array} \begin{array}{c} \text{F M C H Sy DL StJ} \\ \begin{pmatrix} 0 & 1 & 0 & 1 & 0 & 0 & 0 \\ 1 & 0 & 0 & 1 & 0 & 0 & 0 \\ 0 & 0 & 0 & 1 & 0 & 0 & 0 \\ 1 & 1 & 1 & 0 & 1 & 1 & 1 \\ 0 & 0 & 0 & 1 & 0 & 0 & 0 \\ 0 & 0 & 0 & 1 & 0 & 0 & 1 \\ 0 & 0 & 0 & 1 & 0 & 1 & 0 \end{pmatrix} \end{array}$$

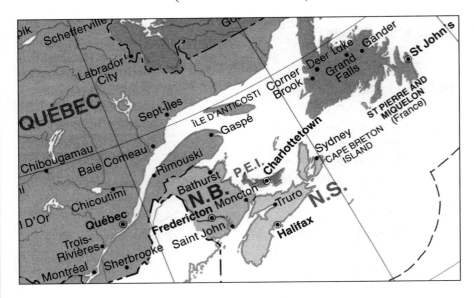

(a) Draw a digraph to represent this information.
(b) Create a matrix that shows the number of one-stop flights between airports.
(c) How many one-stop flights are there from Charlottetown to Halifax? How can you tell by looking at your digraph? At your squared adjacency matrix?

22. The digraph in the margin represents information about city transit. Each vertex is a transfer point.
    (a) Construct the adjacency matrix. Create a title for it that describes its information to transit users.
    (b) Find the number of routes requiring two transfers for passengers traveling from Point D to Point A. Explain what you did.

Transit Routes

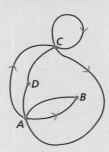

Chapter Project

Designing a Park

Section 2.3 Connections

(a) Draw a digraph that represents a network of five bike trails in a park.

(b) Create an adjacency matrix to represent this digraph.

(c) What information is represented by the square of this adjacency matrix?

(d) Create the squared adjacency matrix.

(e) Select an element in the squared adjacency matrix and describe how it relates to the digraph.

(f) This graph represents a network of bike trails. Represent this information in a row matrix.

Bike Trails

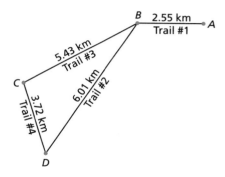

(g) Abby rides the trails shown in part (f) regularly. She travels at a speed of:

- 1 min per km when the trails are in excellent condition;
- 1.05 min per km when the trails are in moderate condition; and
- 1.15 min per km when the trails are in poor condition.

Represent this information in a column matrix.

(h) Use the matrices from parts (f) and (g) to create a product matrix that organizes information about how long it takes Abby to travel each trail in each type of condition.

PUTTING IT TOGETHER

CASE STUDY 1

Suppose that you are booking concerts for a band's tour of Atlantic Canada. The band, based in Toronto, has five days to perform in five cities. They would like to perform the evening of the day they leave Toronto and to return the day after their last concert.

(a) Decide on the five locations for the concerts. Explain your choices.

(b) Determine the most appropriate order for the concerts. Explain your choices.

(c) Determine the band's route and their method of transportation for this route.

(d) Write a report on the band's finances for their Atlantic Tour. This report should include information such as:

Expenses:
- Transportation
- Accommodation
- Meals
- Publicity
- Facility rental
- Other; for example, equipment repairs

Revenue:
- Ticket revenue
- Other; for example, CD and T-shirt sales

CASE STUDY 2

Below is a map of a holiday village. The lines represent walkways and the dots represent condominiums.

The holiday village is still under development. Rebecca and Ken note that it could use some tea rooms.

At first, Ken and Rebecca think of renting space at each condominium, but this would be too expensive. They then consider setting up only one tea room, but realize that many of the residents would not have easy access to that tea room.

Through a survey, Ken and Rebecca determine that most residents would be willing to travel a maximum of two walkways to a tea room.

Now they need to determine the number and locations of the tea rooms.

(a) Determine the minimum number of tea rooms needed. Locate these tea rooms on a map of the holiday village.

(b) Explain the strategies you used to decide where to put the tea rooms. How did you make sure that no condominium was too far away from a tea room?

(c) Do you think it is possible to arrange the tea rooms in a different way so that Ken and Rebecca won't need as many of them?

(d) Construct your own map of a community and pose a similar problem. Describe the problem and its solution.

CASE STUDY 3

The connecting dots and lines of graphs make the ideas in graph theory useful for designing computer systems. For example, imagine that you have been assigned the job of linking twelve computers. Each computer must be linked to all the others. Suppose that each is directly connected to only three others, and that any message that has to pass through more than two computers will not be efficient.

(a) How will you link these computers? Clearly label a diagram showing the appropriate connections.

(b) To solve this problem, you constructed a graph that was *3-regular* and that had a *diameter* of three or less. Research and explain this terminology.

Suppose that you have been assigned a task designing the electrical impulses on a silicon chip. You are allowed to use only one side of the chip and, if any of the connections cross, the chip will short circuit.

(c) Use twelve vertices connected in such a way that each is linked to all the others, but only directly to three others. Each vertex should be linked to every other vertex through a maximum of two other vertices.

(d) To solve this problem, you constructed a graph that was *planar* as well as *3-regular* and that had a *diameter* of three or less. Research and explain this terminology.

EXTENSION 1

Sprouts is a game for two that involves creating network graphs. Begin with two vertices. The first player connects the two vertices by drawing an edge or loop at one of the vertices and then creates a new vertex on the midpoint of that edge. The other player then connects two vertices with an edge or draws a loop and then creates a new vertex at the midpoint of that edge.

There are only two rules:

- No more than three edges can meet at a vertex.
- No edge can cross another edge.

The winner is the last player to draw an edge.

For example,

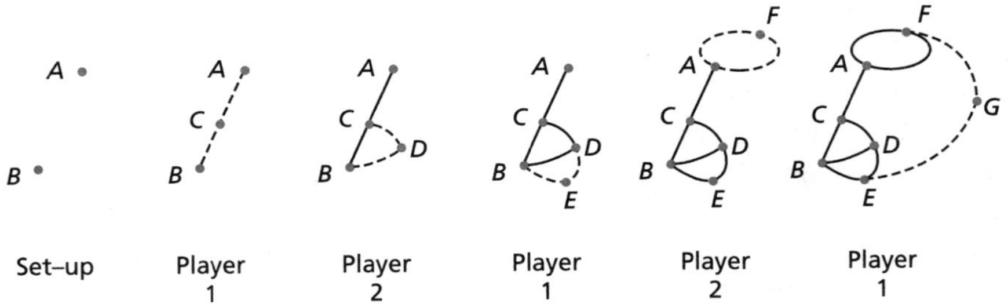

| Set-up | Player 1 | Player 2 | Player 1 | Player 2 | Player 1 |

The game shown here is over, and Player 1 wins, because after five turns, there are three edges at each vertex except for G, so G cannot be connected to another vertex and a loop at G would create a fourth edge.

86 Chapter 2 *Networks and Matrices*

EXTENSION 2

The following graph is a 2-D representation of a cube.

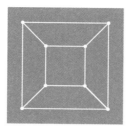

(a) Describe how it can represent a cube. What other 3-D shape could it represent?

(b) Can all the edges of a cube be traveled exactly once? Use the graph to find out.

(c) Draw graphs to represent other 3-D shapes. Which shapes can be traveled without repeating an edge?

EXTENSION 3

Graphs can be used to represent competitions. In round-robin tournaments, each player plays each other player exactly once. Each vertex in this graph represents a player and each edge represents a game that has been played between two players.

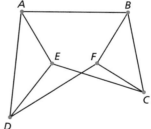

(a) Is the tournament finished? Explain.

(b) Create a graph that represents the same information with the vertices arranged differently.

(c) Create a problem about the tournament that can be answered using the graph.

(d) What other information could be added to the graph to make it more useful?

(e) What will the graph look like when the round-robin tournament is finished?

REVIEW

Key Terms

	page
adjacency matrix	68
digraph	63
dimensions	68
edge	57
element	68
even vertex	58
graph	56
major diagonal	71
matrix	67
network	56
network graph	57
odd vertex	58
product matrix	77
square matrix	79

You Will Be Expected To

- model real-world network situations with appropriate graphs and digraphs.
- interpret network graphs and digraphs to solve related problems.
- represent digraphs with adjacency matrices and vice versa.
- multiply matrices in order to solve related problems.

Summary of Key Concepts

2.1 Creating and Traveling Network Graphs

A *network* is a set of people, places, objects, or ideas that are connected in some way. Networks can be represented in diagrams called *graphs*. The graphs make it easier to understand the network and to solve related problems.

Example

A network of roads between towns patrolled by regional police is represented by the following graph. It is a connected graph because each of the four *vertices* is connected to at least one other vertex by at least one *edge*.

The two edges from vertex A to vertex B might mean that there are two different roads connecting those two towns. It could also mean that there is one road between those two towns but it must be patrolled on both sides.

Graph 2 represents the same police-patrol network but the vertices are arranged differently.

Graph 1 — Network of Roads to be Patrolled

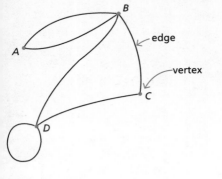

Graph 2 — Network of Roads to be Patrolled

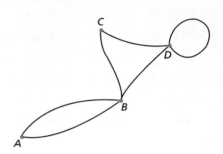

88   Chapter 2 *Networks and Matrices*

Traveling Network Graphs

There are many network problems that can be solved by creating graphs and then finding routes to travel the graphs as efficiently as possible.

Graph 3

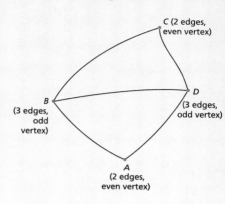

Example

Suppose the police want to travel through the network each night without repeating a road. This is possible for Graph 1 because all the vertices in the network are *even vertices*. A possible route through Graph 1 is A, B, C, D, D, B, A. (Other routes are possible.)

If there were exactly two *odd vertices* in a graph, the police could patrol each road exactly once, starting and ending at different vertices as in Graph 3.

Possible route: B, C, D, B, A, D. (Other paths are possible.)

2.2 Digraphs and Adjacency Matrices

Digraphs are graphs that have edges that indicate the direction of travel.

Example

In this digraph, the single edges connecting pairs of vertices indicate that those roads are one way and can be patrolled in one pass, but only in a specific direction. The double edges indicate that those roads are two way and must be patrolled in both directions.

Digraph 1 — Network of Downtown Streets To Be Patrolled

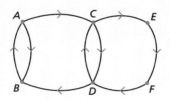

Simplifying Digraphs

Digraph 1 can be simplified. If there are an equal number of edges going in both directions between two vertices, the arrow can be dropped and a number can be used to indicate the number of edges traveling in each direction.

Example
Traveling Digraphs

There are many network problems that can be solved by creating digraphs and then finding routes that travel the graphs as efficiently as possible.

Digraph 1 — Network of Downtown Streets To Be Patrolled (Simplified)

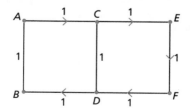

Example

A non-repeating route through Digraph 1, where every road is patrolled exactly once, is not possible because there are two odd vertices.

A non-repeating route through Digraph 1 is possible if it begins and ends at different vertices. This is possible because:

- there are exactly two odd vertices; and
- there are the same number of edges going into each even vertex as there are coming out.

Possible route: A, C, D, C, E, F, D, B, A, B. (Other routes are possible.)

Review 89

Adjacency Matrices

An *adjacency matrix* is a rectangular array of numbers that represents connections between vertices in a network graph.

Example
The adjacency matrix for Digraph 1 is:

Matrix 1

$$
\text{FROM} \quad \begin{array}{c} \\ A \\ B \\ C \\ D \\ E \\ F \end{array} \overset{\displaystyle \text{TO}}{\begin{pmatrix} A & B & C & D & E & F \\ 0 & 1 & 1 & 0 & 0 & 0 \\ 1 & 0 & 0 & 0 & 0 & 0 \\ 0 & 0 & 0 & 1 & 1 & 0 \\ 0 & 1 & 1 & 0 & 0 & 0 \\ 0 & 0 & 0 & 0 & 0 & 1 \\ 0 & 0 & 0 & 1 & 0 & 0 \end{pmatrix}}
$$

The *element* "1" in row 1, column 2 indicates that there is one edge from vertex A to vertex B. The element "0" in row 4, column 1 indicates that there are no edges from vertex D to vertex A.

The *dimensions* of a matrix are described as the number of rows by the number of columns. In the case of Matrix 1, the dimensions are six by six.

2.3 Matrix Multiplication

Two related matrices can be multiplied to create a product matrix with new, related information.

Keep in mind:

- The number and meaning of the columns of the first matrix must be the same as the number and meaning of the rows of the second matrix.
- The product matrix has the row headings of the first matrix and the column headings of the second matrix.
- If the first matrix is m by n and the second matrix is n by p, the product matrix will be m by p.
- Each element in the product matrix is calculated by multiplying each element in a given row in the first matrix by each element in the corresponding column of the second matrix, and then adding the products.

Example

	First Matrix		Second Matrix		Product Matrix

$$\begin{array}{c} \\ X \\ Y \\ Z \end{array}\begin{pmatrix} M & N & O & P \\ 1 & 2 & 3 & 4 \\ 5 & 6 & 7 & 8 \\ 9 & 10 & 11 & 12 \end{pmatrix} \times \begin{array}{c} \\ M \\ N \\ O \\ P \end{array}\begin{pmatrix} A & B \\ 13 & 17 \\ 14 & 18 \\ 15 & 19 \\ 16 & 20 \end{pmatrix} = \begin{array}{c} \\ X \\ Y \\ Z \end{array}\begin{pmatrix} A & B \\ 150 & 190 \\ 382 & 486 \\ 614 & 782 \end{pmatrix}$$

The element "150" in *row 1, column 1 of the product matrix* is calculated using the elements in *row 1 of the first matrix* and *column 1 of the second matrix*:
$1 \times 13 + 2 \times 14 + 3 \times 15 + 4 \times 16 = 150$.

Dimensions:

First Matrix		Second Matrix		Product Matrix
(3 by 4)	×	(4 by 2)	=	(3 by 2)

Squaring an Adjacency Matrix

Multiplying an adjacency matrix by itself, in other words, *squaring a matrix*, results in a matrix in which each element represents the number of connections between pairs of vertices that go through another vertex.

Example

Matrix A represents the number of direct bus connections between cities in a network of four cities.

Matrix A^2 represents the number of connections between pairs of cities in the same network that stop at another city along the way.

$$\begin{array}{c}\text{Matrix } A \\ \begin{pmatrix} 0 & 0 & 1 & 2 \\ 0 & 0 & 2 & 0 \\ 1 & 2 & 0 & 1 \\ 2 & 0 & 0 & 0 \end{pmatrix}\end{array} \times \begin{array}{c}\text{Matrix } A \\ \begin{pmatrix} 0 & 0 & 1 & 2 \\ 0 & 0 & 2 & 0 \\ 1 & 2 & 0 & 1 \\ 2 & 0 & 0 & 0 \end{pmatrix}\end{array} = \begin{array}{c}\text{Matrix } A^2 \\ \begin{pmatrix} 5 & 2 & 0 & 1 \\ 2 & 4 & 0 & 2 \\ 2 & 0 & 5 & 2 \\ 0 & 0 & 2 & 4 \end{pmatrix}\end{array}$$

The element "5" in row 3, column 3 of Matrix A^2 indicates that there are five ways that you can travel from C back to C, with one stop along the way.

If you were to multiply Matrix A^2 by A, in other words, to cube the matrix to create Matrix A^3, you would get an adjacency matrix of two-stop routes.

PRACTICE

2.1 Creating and Traveling Network Graphs

1. The diagram below shows the phone conversations among a group of friends one evening.

   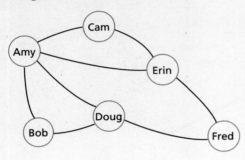

   (a) Explain why this is a network.
   (b) Create a graph of the network.
   (c) What do the vertices represent?
   (d) What do the edges represent?
   (e) Create and solve a problem that could be solved using this graph.

2. Each diagram shows a hiking trail to be cleared. Which trails can be cleared without repeating a section of the trail? Explain.

   Trail 1 **Trail 2**

   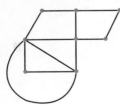

   Trail 3

3. The heavy lines in this street map indicate where parking meters are located along the sidewalks in the downtown area.

   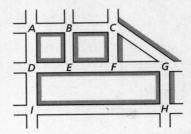

   Can all the parking meters in the downtown area be checked without repeating any sidewalks? Explain why or why not.

4. Leonhard Euler could see how mathematics played a role in everyday life. In the old German town of Königsberg, it was a tradition for people to take a stroll, intending to cross each of the seven bridges exactly once. Euler studied the problem to see if it was possible.
   (a) What did he discover? Explain.

   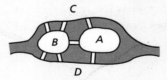

   Bridges of Königsberg

   (b) Since Euler's time, two more bridges have been built, one connecting the south mainland, D, to island A, and another connecting the north and south sides of the river, just west of island B. Is it now possible to travel a route crossing all nine bridges exactly once? Explain.

5. A garbage truck can pick up garbage on both sides of a street in one trip. Use the downtown area shown in Question 3.
   (a) Draw a graph to represent the network that the garbage truck must service.

92 Chapter 2 *Networks and Matrices*

(b) Is it possible to pick up garbage, beginning and ending at the same intersection, traveling each street exactly once?

2.2 Digraphs and Adjacency Matrices

6. Snowplow Network

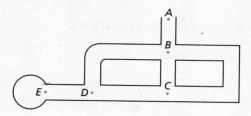

Snowplow Route Digraph

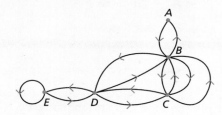

(a) Describe what the digraph tells you about how a snowplow plows a road.

(b) Can a plow travel each edge of the graph exactly once, starting and ending at vertex A? If so, describe the route and explain why it is possible.

7. Simplify the snowplow digraph in Question 6.

8. (a) Create an adjacency matrix to represent the snowplow digraph in Question 6.

(b) How can you use the adjacency matrix to determine the number of directions the snowplow can go when leaving an intersection? The number of ways the snowplow can approach an intersection?

9. Create a digraph for the following adjacency matrix. Simplify the digraph, if necessary.

$$\begin{array}{c} \\ A \\ B \\ C \\ D \end{array} \begin{array}{c} A \ B \ C \ D \\ \begin{pmatrix} 0 & 2 & 4 & 0 \\ 2 & 1 & 0 & 3 \\ 4 & 0 & 0 & 0 \\ 0 & 1 & 0 & 0 \end{pmatrix} \end{array}$$

2.3 Matrix Multiplication

10. (a) Which pairs of matrices can be multiplied?

$$A \qquad\qquad B$$
$$\begin{pmatrix} 3 & 4 \\ 2 & 1 \end{pmatrix} \qquad \begin{pmatrix} 6 & 1 & 0 \\ 4 & 3 & 1 \\ 2 & 1 & 2 \end{pmatrix}$$

$$C \qquad\qquad D$$
$$\begin{pmatrix} 6 & 1 & 6 & 1 \\ 5 & 5 & 0 & 0 \\ 3 & 3 & 1 & 1 \end{pmatrix} \qquad \begin{pmatrix} 8 & 2 \\ 7 & 2 \\ 4 & 6 \end{pmatrix}$$

(b) For each pair of matrices that can be multiplied, create the product matrix.

11. (a) Which matrices in Question 10 can be squared? Explain why.

(b) For those matrices that can be squared, create the product matrix.

12. This digraph shows direct flights among three locations.

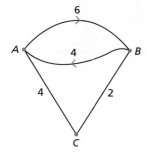

Create a matrix showing the number of flights with one stopover between the locations.

13. This digraph represents an informal communication network among a group of classmates.

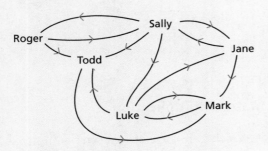

(a) Will Luke hear something directly from Sally?
(b) Will Sally hear something directly from Luke?
(c) Represent the digraph using an adjacency matrix, A.
(d) What will the elements of A^2 mean?
(e) (i) What is the element in the 2nd row, 3rd column of A^2?
   (ii) What does it represent?
   (iii) How does it relate to the digraph?

14. Information about four city school districts has been collected.

Student Transportation (by %)

District	A	B	C	D
Bused	20	30	10	5
Not Bused	80	70	90	95

Student Enrolment

District	Last Year	This Year	Next Year (Projected)
A	8000	8200	8400
B	10 000	10 500	10 800
C	6000	5900	5700
D	7500	8200	9000

(a) Explain why these two matrices can be multiplied.
(b) Multiply the matrices and explain the meaning of the product matrix.
(c) How might the product matrix be of use to city planners?

15. Explain how you can use the headings of two matrices to predict
   (a) whether a product matrix is possible;
   (b) the meaning of the product matrix; and
   (c) if the product matrix will be meaningful.

Chapter Three
Patterns, Relations, Equations, and Predictions

When fictional detectives, such as Sherlock Holmes and Hercule Poirot, are solving crimes, they look for patterns that can be extended to make predictions. These patterns help them to eliminate suspects and narrow possible conclusions. Real-life detectives also look for patterns to help them solve their cases.

In this chapter, you will be studying patterns in different situations that can lead to conclusions and help you make predictions.

After successfully completing this chapter, you will be expected to:

1. Model situations using concrete materials, patterns, and equations.
2. Solve equations presented in many different forms.
3. Create an equation for a line or curve.
4. Make predictions using the lines and curves that you discover.
5. Interpret the values represented by key points on the line or curve.

3.1 Describing Patterns

modeling — the technique of producing a mathematical description or model that can be used to solve a practical problem

Patterns can often be found by making a **model** of a situation. Models can be scale models, tables of values, graphs, or algebraic equations.

Investigation 1
Finding a Pattern

Purpose
In this Investigation you will use a pattern to make predictions about **trains** constructed from cubes. You will also find the number of faces showing when 200 cubes are used to make a train.

train — a number of cubes joined in a row. A train of 3 cubes is shown.

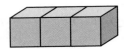

Procedure

A. Place one cube on your desk. How many **faces** are visible?

B. Use two cubes to make a train on your desk. How many faces are visible?

face — one side of a cube

C. Create more trains. Each train will have one cube more than the previous train. Record the number of visible faces on each train.

Number of Cubes	1	2	3	4	5	6	7
Number of Visible Faces	5	8	11	14	17	20	23

sequence — a set of numbers arranged in order according to a pattern or rule. For example, 2, 4, 6, 8, . . . is a sequence of even numbers.

What pattern do you see in the **sequence** of numbers you collected?

Investigation Questions

— Note —
In a sequence, there is a first term, a second term, and so on.

1. How many visible faces are there for a train of 11 cubes? Test your answer.

2. What is the number of visible faces for a train of 12 cubes and for a train of 15 cubes? How confident are you in your answers? Why?

3. List restrictions on possible values for the number of visible faces.

4. Predict the number of visible faces for a train of 200 cubes.
   - Describe how you made your prediction.
   - How confident are you in your prediction?
   - How could you check your prediction? Is this practical?

96 Chapter 3 *Patterns, Relations, Equations, and Predictions*

FOCUS A: Graphing Results and Making Predictions

Suppose you decided to graph the ordered pairs from the Investigation to find the number of visible faces for a train of with a given number of cubes.

A. Copy this coordinate axis. Graph the table of values from Investigation 1.

B. Does it make sense to join the points on the graph with a line or a curve? Determine whether the data are **continuous** or **discrete** to help you decide.

C. What numbers are on the *c*-axis? What numbers are on the *f*-axis? These numbers are part of the **domain** and the **range**.

D. Describe the patterns shown in the graph. Discuss:
   (a) the geometric pattern formed by the data points.
   (b) how the graph displays patterns from the table of values.
   (c) why all the points of the graph are in the first quadrant.

Focus Questions

5. (a) Use your graph to find the number of visible faces for a train with 18 cubes.

   (b) Explain how you found your answer. Is a graph the best way to do this? Explain.

   (c) How confident are you that your answer is correct?

6. What is the most reasonable way to find the number of visible faces for a train of 200 cubes? Explain.

7. Is it easier to use a graph or a table of values to make predictions for a large number of cubes? Explain.

Think about...

Step A
Explain why it makes sense to graph the number of cubes as the independent variable and the number of visible faces as the dependent variable.

continuous data — an infinite number of values exist between any two other values in the table of values or on the graph. Data points are joined.

discrete data — a finite number of data points exist between any two other values

domain — the set of all possible values for the independent variable in any relation

range — the set of all possible values for the dependent variable in any relation

Domain and range can be expressed in set notation as follows:
$D = \{x | x > 0, x \in N\}$
$R = \{y | y > 0, y \in N\}$

In this example, the values in both the domain and the range belong to the set of natural numbers.

— Note —
Any coordinate grid has four quadrants as shown. In this problem, we use only quadrant 1.

2	1
3	4

Think about...

Question 8

A complex problem can often be solved by looking at a simpler version of the same problem. The simpler version may provide a pattern you can use. Explain how this applies.

—Note—

When answering part 8(d), think about the domain and the range of the graphs.

8. Suppose you were asked to find the number of visible faces for a train with 1000 cubes.

   (a) Explain why it would not be practical to use a cube model, table of values, or graph to find the number of visible faces.

   (b) Create a formula that describes the relationship between the number of visible faces and the number of cubes used.

   (c) Describe how you could check the accuracy of your formula. This will tell you how well it models the relationship.

   (d) Graph your formula using a graphing calculator. Describe all the ways that your hand-drawn graph will differ from that produced using technology. Explain why these differences occur.

Check Your Understanding

9. Look at the pattern in the trains in Investigation 1 and Focus A. Find the number of visible faces for each of the following trains.

   (a) 22 cubes (b) 30 cubes (c) 40 cubes
   (d) 50 cubes (e) 60 cubes (f) 70 cubes

Investigation 2
Using Patterns and Models

Purpose

Investigate the pattern in the sequence of crosses. Find out whether a cross can be made that uses 37 cubes.

Cubes are arranged to form this sequence of crosses.

—Note—

In this pattern, consider a single cube as a cross. The single cube is the first cross.

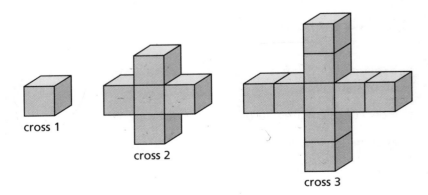

cross 1   cross 2   cross 3

98   Chapter 3 *Patterns, Relations, Equations, and Predictions*

A. Describe the pattern that connects the cross number to the number of cubes.
   - Draw a graph of this relationship.
   - Explain how the pattern shows up in the graph.
   - Write a formula that will allow you to find the number of cubes when the cross number is known. Explain how you found this formula.

B. Describe the pattern that connects the number of cubes to the number of exterior faces.
   - Draw a graph of this relationship.
   - Explain how the pattern shows up in the graph.
   - Write a formula that will allow you to find the number of exterior faces when the number of cubes is known. Explain how you found this formula.

C. Describe the pattern that connects the cross number to the number of exterior faces.
   - Draw a graph of this relationship.
   - Explain how the pattern shows up in the graph.
   - Write a formula that will allow you to find the number of exterior faces when the cross number is known. Explain how you found this formula.

D. Explain whether or not it is possible to build a cross using exactly 37 cubes. Find the number of exterior faces for this cross.

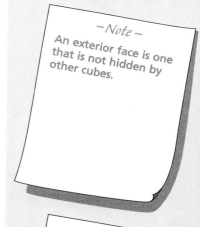

—Note—
An exterior face is one that is not hidden by other cubes.

—Note—
When you use a graph to find a value, you can only estimate the number of cubes. If you need an exact answer, you have to use another method like an expression or an equation.

Investigation Questions

10. For each graph in Steps A, B, and C:
    (a) Are the data continuous or discrete? How do you know?
    (b) What is the independent variable? The dependent variable?
    (c) Describe the domain and the range.

11. Which number of cubes could be used to make a cross in the sequence? Which could not?
    (a) 45 cubes   (b) 50 cubes   (c) 118 cubes   (d) 120 cubes

12. Think about how you have used the strategies below to model the patterns you found when making crosses with different numbers of cubes. For what numbers is each strategy a reasonable choice?
    (a) building a model with cubes
    (b) using a table of values
    (c) drawing a graph
    (d) looking for a pattern in the sequence of values
    (e) finding a formula

Check Your Understanding

Question 13
On which graph is it easiest to identify a particular pattern?

13. Use the graphs that follow.
    (a) Name the coordinates of the points shown on each graph.
    (b) Name another point on each graph, that is not marked, but fits the pattern.
    (c) How is a linear graph like a non-linear graph? How is it different?
    (d) Find the domain and range for each graph. Include the data you added in part (b).
    (e) Are the data discrete or continuous?

A
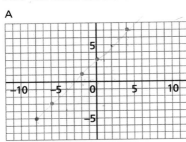

B
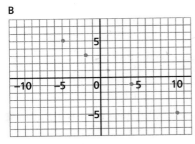

$(-6,-3)$
$(-2, 1)$
$(0, 3)$
$(4, 7)$

C
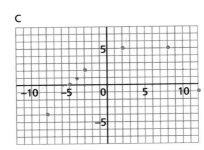

D
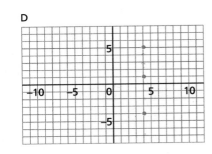

14. Frau was driving a motorcycle. She recorded the gas consumed each hour of her trip.

Time (h)	0	1	2	3	4	5
Gas Consumed (L)	0	0.5	1.0	1.5	2.0	2.5

Question 14
Is the graph continuous or discrete? Explain.

(a) Describe the pattern in the table of values.
(b) Sketch a graph for the data.
(c) Describe how the pattern in (a) shows up in the graph.
(d) Predict the amount of gas that the motorcycle would consume in 6 h. Explain how you found your answer.
(e) Write the domain and range for your relationship.

15. (a) Group pennies in stacks of 4, 8, 12, 16, and 20. Measure and record the height of each stack. Graph the data.
    (b) Describe how the appearance of this graph differs from the graph in Question 14.

16. A photographer charges a sitting fee of $20 and $6 for each photograph ordered.
    (a) Find the total cost for ten photos. Did you use a graph, an equation, or something else? Give reasons for your choice.
    (b) A customer has $100. How many photos can be ordered?

17. A new theatre has 20 seats in the first row, 23 in the second row, 26 in the third row, and so on. Find the number of seats in row 12. Explain how you found your answer. What is the practical limit to the number of seats based on this pattern? Justify your answer.

18. An Internet provider charges $20.00 per month plus $2.00 for each hour of use. After how many hours would you be charged $60.00 for the use of the Internet? Show all of your work.

Investigation 3   20hrs
Connecting Patterns with Equations

Purpose
Solve a problem about Internet use by graphing.

After researching the rates of different Internet providers, Daniella decided to use Company A. Company A charges $20.00 per month plus $2.00 per hour. Daniella's first monthly bill was $80.00. How many hours was she on-line in the first month?

Procedure

A. Use the relationship to construct a continuous graph for Company A's plan. How is the graph's appearance affected by the domain and the range?

B. Explain how you could use the graph to solve the problem.
   • Use the graph to estimate the answer.
   • Discuss how you could check the accuracy of your estimate.

C. Write a formula to find the cost of Internet use if the number of hours used in one month is known.

   Write an **equation** to show the number of hours of Internet use for Daniella's bill of $80.00. Explain how you could use this equation to solve the problem.

Think about...

Your predictions
Is it easier to use the table of values, the graph, or the pattern to make the prediction? Why?

Think about...

Question 16(b)
Why does the customer not spend the entire $100?

Think about...

Step A
Explain why using a table of values is not a very efficient way of solving this problem.
Describe the domain and the range for this relationship.

Step B
Discuss the effectiveness of using the graph to solve the problem.

Step C
Explain why using a formula might be the best strategy for solving this problem.

equation — a mathematical sentence showing that expressions are equal in value. For example, $x + 2 = 10$. When equations are written, letters are used to represent the variables.

Check Your Understanding

— Note —
In Question 19, assume that each partial kilometre driven results in a partial charge.

19. Heather and her friends take a taxi to a Great Big Sea concert.
    - The taxi meter starts at $2.50.
    - Each kilometre driven costs $2.00.
    - Heather pays $31.50 for the taxi ride.

    How far did the taxi travel?
    (a) Draw a graph. Use it to estimate the solution. Check the accuracy of your estimate.
    (b) Explain why the equation $31.50 = 2.5 + 2d$ accurately represents this problem.
    (c) Describe a method you can use to get an exact solution.

20. Lakshmi and Vasim are having a party to celebrate their daughter's success in the kayaking event at the Canada Games. A DJ for the party charges $200 plus $50 per hour.
    (a) Write the equation that represents this relationship.
    (b) For how long will the DJ play for $525?

Think about...

Questions 21 and 22

How might you determine the number of CDs sold that will result in Craig and Sonia earning the same amount of money? Find this amount.

21. Sonia signed a contract with a recording company. She receives $10 000 cash on signing and $2.00 for each CD sold.
    (a) Suppose Sonia's CD sold 50 000 copies. How much will she earn?
    (b) Find the number of CDs sold if she earns $40 000.

— Note —
In Questions 23 and 24, assume that each partial kilometre driven results in a partial charge.

22. Craig signed a contract with a different CD company. He gets no signing bonus and $2.50 for each CD sold.
    (a) Describe the domain and range.
    (b) Suppose Craig's CD sold 50 000 copies. How much will he earn?
    (c) Find the number of CDs sold if he earns $40 000.

23. A taxi company charges a basic fee of $5 plus $0.30 per kilometre travelled.
    (a) What is the dependent variable? What is the independent variable?
    (b) Describe the domain and range.
    (c) What is the distance traveled for a charge of $32?

Think about...

Questions 23 and 24

For what distance traveled would taxis from the two companies charge the same amount of money?

24. Another taxi company charges a basic fee of $3 and $0.50 per kilometre.
    (a) What is the distance traveled for a ride that costs $32?
    (b) Is this taxi company more expensive or less expensive than the one in Question 23 for a ride of 15 km? For a ride from your house to school?

102   Chapter 3 *Patterns, Relations, Equations, and Predictions*

25. The following graphs represent possible situations modeled by equations.
    (a) Describe a situation that each graph could represent.
    (b) Make up a problem based on the situation. Have others solve your problem.
    (c) Explain why it does or does not make sense to extend the graph beyond the values provided.

A

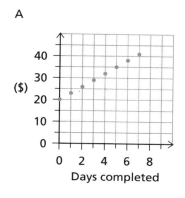

B

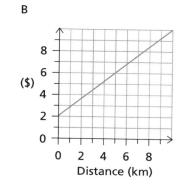

Chapter Project

Designing a Dog Pen

Seana, a home owner near Centennial Park in Moncton, N.B., needs to construct a rectangular outdoor pen for her dog. She can buy a maximum of 24 m of chain-link fencing.

(a) Sketch a number of pens that can be constructed using 24 m of fencing.
(b) Record the length and width of each pen in a table of values.
(c) Plot a graph to show the relationship between width and length. Use width as the independent variable.
(d) Describe the relationship between length and width.
(e) Express this relationship as an equation.
(f) What other factors might affect Seana's construction plans?

3.1 Describing Patterns 103

3.2 Solving Problems by Solving Equations

solving an equation — finding the value of the variable that makes the equation true

Finding a good representation or model is often the key to solving a problem. When a problem involves a relationship that can be written algebraically, that usually leads to solving an equation.

FOCUS B: Solving Equations with Algebra Tiles

Purpose
This Focus reviews how algebra tiles can be used to solve equations like the one you developed to solve the Internet user problem in Investigation 3.

Example 1
Solve $2 - 3x = 5$.

Solution
Step 1 Model the equation using tiles.

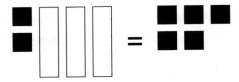

$$2 - 3x = 5$$

Step 2 To isolate the term containing the variable, use an appropriate number of negative unit tiles to make a net result of zero unit tiles on one side.

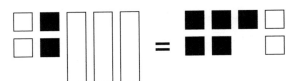

$$2 - 2 - 3x = 5 - 2$$
$$\text{or } 2 + (-2) - 3x = 5 + (-2)$$

104 Chapter 3 *Patterns, Relations, Equations, and Predictions*

Step 3 Simplify.

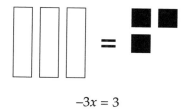

$-3x = 3$

Step 4 Divide both sides by –1.

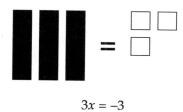

$3x = -3$

Step 5 Isolate the variable.

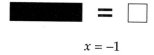

$x = -1$

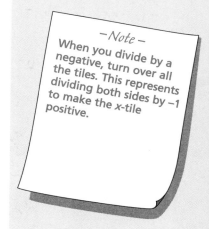

— Note —
When you divide by a negative, turn over all the tiles. This represents dividing both sides by –1 to make the *x*-tile positive.

Focus Question

1. Refer to Example 1.
   (a) In Step 2, why were negative tiles added to the left side?
   (b) Why were the negative tiles added to the right side as well?
   (c) Explain how Steps 2 and 3 are related by the zero property. Then explain how you went from Step 2 to Step 3.
   (d) Explain what happened when you went from Step 4 to Step 5.

Check Your Understanding

2. The equation $3x + 2 = 5$ can be solved using algebra tiles.
   - The steps of the solution are shown using tiles in A, B, C, and D.
   - The algebra steps are shown in (a), (b), (c), and (d).

   Write out the algebraic steps in order. Then match each tile diagram to the corresponding algebra step.

   (a) $3x + 2 = 5$
   (b) $x = 1$
   (c) $3x = 3$
   (d) $3x + 2 - 2 = 5 - 2$

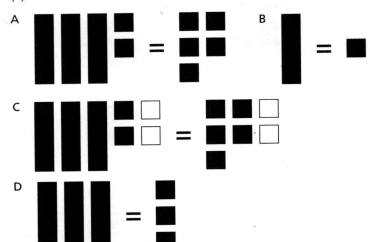

3. Solve each equation. Determine which equation has the largest solution value.

   (a) $3x + 2 = 8$
   (b) $4 = 2(x - 5)$
   (c) $3x - 7 = -1$
   (d) $4x + 3.5 = 7.5$

4. For each equation, state whether or not algebra tiles could be used to solve the equation efficiently.
   - Explain why in each case.
   - Solve those where the tiles would be efficient.

   (a) $2(m + 3) = 22$
   (b) $3y - 4 = 2$
   (c) $5 - x = 9$
   (d) $-4 - 5x = 15$
   (e) $-9 = 20 - 5n$
   (f) $3k + 16 = \frac{8}{3}$
   (g) $12 = 7z - 9$
   (h) $15 = 30 - 6h$
   (i) $\frac{2}{3}(x - 2) + \frac{1}{2}(x - 2) = 1$
   (j) $\frac{3}{5}(y + 2) - \frac{2}{3}(y - 1) = \frac{1}{4}$

— Note —
In Question 4, the equations that can't be solved using tiles will be revisited later in the chapter.

106 Chapter 3 *Patterns, Relations, Equations, and Predictions*

Extending Equation Solving

Sometimes, you may be asked to solve an equation when you don't have enough tiles. Other times, fractional tiles may be needed but fractional tiles do not exist. You need to extend or generalize the equation-solving process to deal with these situations.

Example 2
Solve $20 + 4p = 60$.   You need 84 tiles. (Probably too many.)

Solution

$20 + 4p = 60$ Subtract 20 from both sides to isolate the
$4p = 40$ term containing the variable.

$\dfrac{4p}{4} = \dfrac{40}{4}$ Divide by 4 to isolate the variable.

$p = 10$

Focus Questions

5. Suppose the original equation had been $20 - 4p = 60$. How would you solve it?

6. Suppose the original equation had been $20 + 4p = 58$. How would you solve it?

Check Your Understanding

7. Solve the equations in Question 4 that could not be solved using algebra tiles. Verify each solution.

8. Solve each equation. Explain why using tiles would not be a good choice in each case.

   (a) $3(m - \dfrac{2}{3}) = 25$

   (b) $-7 = 8 - 4x$

   (c) $12x + 23 = 39$

   (d) $2.1x + 4.3 = 5.2$

   (e) $\dfrac{2}{3}x + \dfrac{1}{4} = \dfrac{3}{5}$

   (f) $\dfrac{3}{4}(x + 1) - \dfrac{1}{5}(x + 2) = 15$

Focus D: Solving Problems

Example 3

The student council rented a bus to take fans to the basketball championship game.
- The basic cost to rent the bus is $300.
- The driver charges an additional $1.50 per kilometre.
- The total bill was $438.

How far did the bus travel?

Solution

First, set up an equation for the total cost of the trip.

Use the variable d to represent the number of kilometres traveled.

total cost = initial value + $1.50 for each kilometre

$$\$438 = \$300 + 1.50d$$

Solve the equation.

$438 = 300 + 1.50d$	To isolate the term containing the variable, subtract 300 from each side.
$138 = 1.50d$	Simplify.
$\dfrac{138}{1.50} = \dfrac{1.50d}{1.50}$	To determine d, divide both sides by 1.50.
$92 = d$	

To check that the solution is correct, substitute your answer in the original equation.

LS   RS
438            $300 + 1.50d$
               $= 300 + 1.50(92)$
               $= 438$

Thus, $d = 92$ is a correct answer.

To check whether the solution to the problem is reasonable, think about the information given in the problem.

92 km × $1.50/km = $138
base cost = $300
Total cost is $300 + $138 or $438.

Example 3
Why can't you use algebra tiles to solve the equation?

Verifying
Is it possible to get a correct solution to the equation but not a correct solution to the problem? Explain.

Verification shows that the solution to the equation is also a reasonable solution to the problem.

You can conclude that the bus did travel a total of 92 km.

Focus Questions

9. Suppose another bus company charges a basic rental fee of $200 plus $2.90 per kilometre. Would traveling with this company for the same trip be less expensive? Explain.

10. Explain why constructing a table of values and/or using a graph would not be good strategies for getting an exact solution for this problem.

Check Your Understanding

For each problem in this section, do the following:
- write the equation that models the situation;
- state what quantity each variable represents; and
- verify each solution.

11. (a) The cost of binding a book is $5.00. There is also a charge of $0.01 per page. The total cost to bind the book is $9.80. How many pages are in the book?

    (b) Suppose the book binder offered to bind the book for $1.00 plus $0.04 per page. What would be the total charge for the book?

    (c) Which choice would be the most economical? Why?

12. The student council is selling tickets to a video dance. The dance will be in two weeks. They sold 80 tickets the first day and know that the average number of tickets sold after that is 30 each day. There are 500 tickets in all. For how many days would you expect tickets to be on sale?

13. The cost of binding a book depends on the number of pages in the book. A printing company provides two options for customers.

    Option 1   Cost = $5.00 + 0.04 × number of pages
    Option 2   Cost = $1.00 + 0.20 × number of pages

    Which option allows for the greatest number of pages if the total cost is

    (a) $40?   (b) $20?   (c) $6?   (d) $5.80?

Think about...

Question 13
Why do you think the set-up costs are different in each option?

Chapter Project

Designing a Dog Pen

Seana uses a perimeter of 24 m to investigate designs for a dog pen.

(a) A friend suggests that she make the pen twice as long as it is wide. Write a perimeter equation for this pen.
(b) What is the length and width of this pen?
(c) What is the length and width of a pen that is 3 times as long as it is wide?
(d) What would the dimensions of the pen be if the back wall of the house is used as a side of the pen?
(e) What would the dimensions of each pen be if two sides of the yard fence are used as sides of the pen?
(f) What are the advantages and disadvantages of each pen?
(g) What do you think Seana should do?

3.3 Decision Making and Patterns

Two Internet providers have gone into business in your rural area. You receive flyers from them. Your home business needs Internet service and you decide to choose one of the two companies.

Company A

$20.00 per month
and $2.00 per hour

Company C

$10.00 per month
and $2.50 per hour

For what number of hours of Internet use are the costs the same?

Investigation 4 ᐸReəd

Making Decisions

Purpose

Choose an Internet provider by developing a strategy that builds upon the skills you developed in Section 3.2.

Procedure

A. Graph the relationships described for Company A and Company C on the same grid.

B. The graphs of the two relationships cross. What do the coordinates of the **intersection point** tell you about the problem?

C. Describe the accuracy of the solution provided by your graphs.

D. Explain why the solution to $20 + 2h = 10 + 2.5h$ would provide an exact solution for this problem.

Investigation Questions

1. Which plan would be less expensive for someone who uses the Internet for about 12 h per month?

2. Which Internet provider do you think your school should select? Why? List any assumptions that you made.

3. After how many hours does Company A become less expensive than Company C?

— *Note* —

In Section 3.2, you saw how a problem could be modeled algebraically. The problem could then be solved by solving an equation. Problem situations often lead to more complicated algebraic models that result in complicated equations.

Think about...

Step A

What assumptions have you made about the domain of these relationships?

intersection point — the point where the graphs of two equations cross

Think about...

Step D

What does h represent in this equation?

— *Note* —

When graphing equations and finding the intersection point, you could use your graphing calculator or spreadsheet programs.

Check Your Understanding

Think about...

Question 4
Why does one graph have a steeper slope than the other? Why are the y-intercepts different?

4. Refer to these graphs.
   (a) Estimate the coordinates of the point of intersection.
   (b) Find the coordinates in (a) exactly by solving an equation.
   (c) Create a problem situation that would be represented by these graphs. Solve the problem.
   (d) Interpret your solution in the context of your problem.

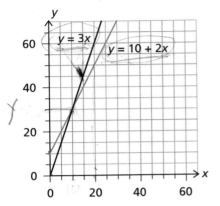

— Note —
You could use graphing technology to solve question 5.

5. Solve each equation by graphing. Explain how you solved each and verify each solution.
   (a) $6x + 3 = 5x + 5$ (b) $4x - 5 = 2x + 3$

6. (a) Create a problem situation that could be represented by the equation in Question 5(a).
   (b) Create a problem situation that could be represented by the equation in Question 5(b).

7. A high-school music group plans to record a demo tape. Two recording studios have provided rental rates.
   (a) Copy and complete each table of values. Then sketch a graph.

Cheap Sound Studio		Rockin' Recordings	
h	$30 + h$	h	$15 + 2.25h$
0		0	
1		1	
2		2	
3		3	
4		4	

Think about...

Question 7
Why does one company have an h and the other have $2.25h$? What effect does this have on the graph?

   (b) What do 30, h, and 15 likely represent in the expressions above?
   (c) Use a graph to solve $30 + h = 15 + 2.25h$. Is your answer exact or an estimate? Explain.
   (d) How does your solution relate to the rental of a recording studio?
   (e) Could these graphs reach outside of quadrant 1? Explain.

8. There are two video stores in your area.
   - Austin Videos charges $5 per video.
   - Babatunde Movie Mania charges a $60 membership fee per year plus $3 per video.

   (a) How many videos would you need to rent in one year for the cost at both stores to be the same?

   (b) Suppose that you rent two movies per month. Which store offers a better deal for you?

   (c) For up to how many rentals is Austin cheaper than Babatunde?

9. (a) Solve $2x + 6 = \frac{2}{3}(9 + 3x)$ by graphing. What do you notice?

   (b) Is $2x + 6 = \frac{2}{3}(9 + 3x)$ an **identity**? Why?

   (c) To what set of numbers does your solution belong? What assumptions have you made?

identity — an equation that is true for all values of the variables involved

Focus E: Extending Equation Solving

In Section 3.2, you generalized solving simple equations so that you didn't need to use algebra tiles. You can do the same for more complicated equations.

– Note –
You would need 72 unit tiles because there are 33 on one side and 39 on the other. You would also need 14 x-tiles.

Example 4
Solve $10x - 33 = 4x + 39$.

Solution
To use tiles, you would need 72 unit tiles. Since this isn't practical, you need to carry out the solution process without using tiles.

$10x - 33 = 4x + 39$	Subtract $4x$ from both sides to remove the variable term from the right side.
$6x - 33 = 39$	
$6x = 72$	Add 33 to both sides.
$\frac{6x}{6} = \frac{72}{6}$	Divide both sides by 6.
$x = 12$	

To solve some equations, you will need to use the **distributive property**.

distributive property — the property of distributing one operation over another without changing the value of the expression

For example,
$3(x + 2) = 3 \cdot x + 3 \cdot 2$
$= 3x + 6$

– Note –

In Example 5, the distributive property for $3(x + 2) = 3x + 6$ can be represented using tiles as follows.

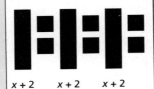

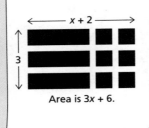

Area is $3x + 6$.

Example 5
Solve $3(x + 2) = 2(x - 1)$.

Solution

$3x + 6 = 2x - 2$ Apply the distributive property.
$3x = 2x - 8$ Subtract 6.
$x = -8$ Subtract $2x$.

Example 6
Solve $5(x - \frac{3}{5}) = x - 4$.

Solution

$5x - \frac{15}{5} = x - 4$ Apply the distributive property.
$4x - 3 = -4$ Subtract x and simplify.
$4x = -1$ Add 3.
$x = -\frac{1}{4}$ Divide by 4.

Check Your Understanding

10. These equations are from Questions 4, 5, 7, and 8 in this section. Solve and verify each. Compare these solutions to your solutions from graphing.
    (a) $3x = 10 + 2x$
    (b) $30 + h = 15 + 2.25h$
    (c) $5v = 3v + 60$
    (d) $6x + 3 = 5x + 5$
    (e) $4x - 5 = 2x + 3$

11. Solve each equation.
    (a) $2n - 5 = n + 3n + 3$
    (b) $4x - 6 = 9 - x$
    (c) $\frac{1}{4}x + \frac{2}{3}x = \frac{1}{4}$
    (d) $\frac{3}{5}y + \frac{1}{2}y = 12$

12. Solve and verify each equation.
    (a) $5(m + 2) - 3(m + 3) = 5(m - 1)$
    (b) $4(m + 2) - 3(m + 3) = 5(m - 1)$
    (c) $-(z + 8) = 5(z - 0.8)$

FOCUS F: Equations with Fractional Expressions

Purpose
Develop a process to solve $\dfrac{3x+2}{6} = \dfrac{x}{6}$.

Procedure

A. Look at the equation above. How are the two expressions alike? How are they different?

B. Brian claimed that he could solve this equation by solving $3x + 2 = x$. Why do you think Brian made this claim? Is he correct?

C. Solve $3x + 2 = x$. How can you decide if this is a solution to the original equation? Is this a solution? Was Brian correct?

Focus Questions

13. Summarize the process Brian used to solve $\dfrac{3x+2}{6} = \dfrac{x}{6}$. Explain why Brian used these steps.

14. Use Brian's process to solve each equation. Verify each solution.
    (a) $\dfrac{x+3}{2} = \dfrac{3x}{2}$
    (b) $\dfrac{5x-1}{4} = \dfrac{2x+5}{4}$
    (c) $\dfrac{3x+1}{3} = \dfrac{2x+4}{3}$

15. Use the same process to try to solve these equations. Verify your solution and summarize the process you used.
    (a) $\dfrac{4x+1}{2} = \dfrac{3x+7}{4}$
    (b) $\dfrac{3x-1}{3} = \dfrac{2x+5}{7}$

Check Your Understanding

16. Explain why $\dfrac{2x+1}{2} + \dfrac{3x-1}{2} = 0$ and $\dfrac{5x}{2} = 0$ have the same solution.

17. Solve these equations.
    (a) $\dfrac{5x-3}{2} = \dfrac{2x-9}{2}$
    (b) $\dfrac{4-3x}{4} = \dfrac{x+1}{4}$
    (c) $-3(x+1) + \dfrac{x+5}{3} = \dfrac{2x-9}{3}$
    (d) $\dfrac{3h-8}{2} - \dfrac{1+2h}{3} = \dfrac{1-3h}{4}$

18. An outdoor swimming pool opens for 14 weeks each year. There are two payment choices:
    - you can pay $3 each visit; or
    - you can buy a season pass for $80 and pay $1 per visit.

    For what number of visits would the cost be the same?

19. Suppose these figures have the same perimeter. Find the value of x.

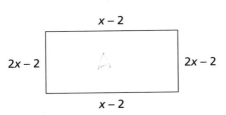
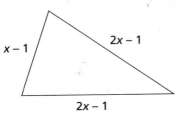

20. Suppose these rectangles have the same area. What is the value of x?

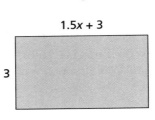
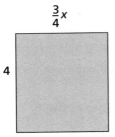

21. Solve these equations.
    (a) $3(x + 2) = 4(x - 1)$
    (b) $5p + 3 = 4(p - 2)$
    (c) $7(q - 3) = 5(q - 2)$
    (d) $4(y - 2) = 3(y + 1)$
    (e) $2(m + \frac{4}{5}) = 7(m - \frac{1}{5})$
    (f) $-(1.1 + x) = 2(x - 2.6)$
    (g) $\frac{1}{2}(2n + 1) = 3(1.5 - n)$
    (h) $-0.5(2t - 4) = 2(\frac{t}{4} + \frac{1}{4})$

Option 1: basic fare of $5 plus 10¢ per kilometre.

Option 2: basic fare of $3 plus 12¢ per kilometre.

22. A taxi company offers two payment options as shown in the margin. For what distance do the two options cost the same?

23. You have $70. Your friend has $148. You plan to save $5 each week and your friend plans to save $3 each week. After how many weeks will you have the same amount of money?

24. Solve these equations.
    (a) $3x + 5 + 2x = x - 7 - 2x$
    (b) $4m + 3m - 9 = 2m + 5m - 8$
    (c) $7 - 3p + 4p = 2 - 9p + 12$
    (d) $4(y - 2) - 9 = 2y + 7 - 3y + 12y$
    (e) $12 - 6a + 2.5a = -9.5a + 7a + 15$
    (f) $-4(x - 3) + x = -2x - \frac{3}{4}$
    (g) $-3(x + 2) + 2(x + 4) = -2(0.5x - 1)$

25. Solve these equations.
    (a) $\frac{x - 2}{3} = \frac{x}{2}$
    (b) $\frac{2x + 6}{7} = \frac{x - 5}{2}$
    (c) $\frac{2}{3}x + 2 = \frac{3}{5}x - 5$
    (d) $\frac{2x + 5}{6} - \frac{2}{3} = \frac{2x}{5}$

116 Chapter 3 Patterns, Relations, Equations, and Predictions

3.4 Predictions and Lines: $y = mx + b$

Suppose you were given only the graphs representing two Internet providers. How might you use the graph to learn more about the monthly charges of each company? How might this help you decide which plan is better for you?

slope y-intercept form — linear equations that are written in the form $y = mx + b$ or $y = b + mx$, where m is the slope and b is the y-intercept

Focus G: Connecting Equations and Graphs of Lines

For the Internet graphs:
- the hourly rate is the value of m. This is the **slope** of the line.
- the initial (base rate) value is b. This is the **y-intercept**.

To find the slope, m:
- find the *run* $(x_2 - x_1)$ as you move from left to right;
- find the *rise* $(y_2 - y_1)$ as you move up and down; and
- find the ratio $\dfrac{\text{rise}}{\text{run}} = \dfrac{y_2 - y_1}{x_2 - x_1}$.

This is the same as finding the average hourly cost.

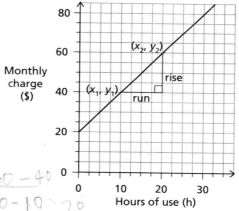

$\dfrac{60-40}{20-10} = \dfrac{20}{10} = 2$

slope — the steepness of a line (growth rate)

y-intercept — the point where a graph crosses the y-axis; the point where $x = 0$ (initial value)

— Note —
When you think of slope, you might think of:
- slope of a driveway
- slope of a wheelchair ramp
- steepness of a hill.

To compute slope, you compare rise and run.
slope = $\dfrac{\text{amount of rise}}{\text{amount of run}}$
 = $\dfrac{y_2 - y_1}{x_2 - x_1}$

Example 1
To deliver soil, Tip Top Soil has a fixed charge plus a charge per "yard" of soil as shown in the graph.
(a) Write an equation for the cost to deliver soil. Use the slope y-intercept form.
(b) How many yards of soil can you purchase for $125?

$(24, 110)$
$(10, 60)$

$\dfrac{110 - 60}{24 - 10}$

$\dfrac{50}{14}$

— Note —
When moving up or down to find the rise, you do not just count the grid squares. Why does counting grid squares not give you the slope?

$125 = 3.55 + b$
$+25 \quad 100 = 3.55$

$y = 3.55 + 25$

Think about...

The Solution

What does c represent?

What does s represent?

Show how the equation was found.

Solution

(a) The equation of the line is $c = 3.5s + 25$.

(b) $125 = 3.5s + 25$

Solve this equation using the skills from previous sections.

Investigation 5
Constructing Graphs and Equations

Purpose

Make decisions about Internet providers using only the graphs of their charge plans.

Procedure

The rates for two Internet companies are shown on the graph.

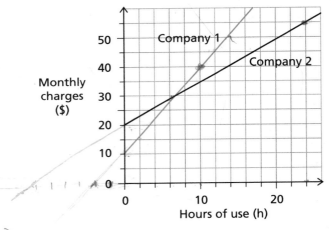

A. Write an equation to represent Company 1's monthly rate.

B. Write an equation to represent Company 2's monthly rate.

C. The monthly rate for Company 3 is graphed below.

Find the equation for this company's rate plan. Describe the plan in words. Write a reasonable domain and range for Company 3.

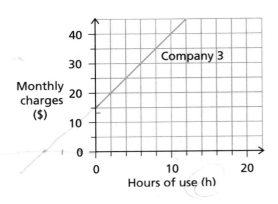

— Note —

In steps A and B, think about what the constant term represents. What do the variable terms represent? Describe each plan in words.

Investigation Questions

1. Refer to the rates for Internet companies 1, 2, and 3. Suppose you have $50.00 per month to spend. Which company might you use? Why? Describe how you made your choice.

2. Refer to the graphs for each equation.
   - The point where the line crosses the c-axis is the c-intercept.
   - The point where the line crosses the h-axis is the h-intercept.

   (a) Estimate the h-intercept and c-intercept for each graph.
   (b) Interpret the meaning of each intercept as it relates to each Internet provider's plan.
   (c) How does each intercept relate to the equation for each plan?

3. From each graph in Investigation 5, read any two points. Each will show hourly cost for a certain length of time.
   (a) Find the average hourly cost using two points from each graph.
   (b) How do these rates relate to the equations for each plan?

4. Suppose you only had the graph of the monthly charges. You did not know the equation or the plan. How could you find the monthly base rate and the hourly rate?

intercept — the point at which the graph crosses each axis

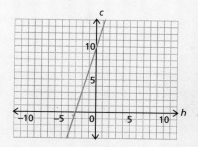

The line crosses the horizontal axis when $c = 0$. The line crosses the vertical axis when $h = 0$.

Think about...

Intercepts in Question 2

Are your intercept values exact or are they estimates? Are estimates good enough for this question?

Check Your Understanding

5. The distance a car travels over time is shown in the graph.
   (a) Discuss the speed of the car during the trip.
   (b) How far from home did the journey begin? How do you know?
   (c) Write an equation to describe the relationship between distance and time.

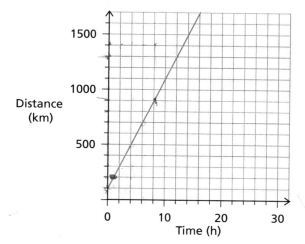

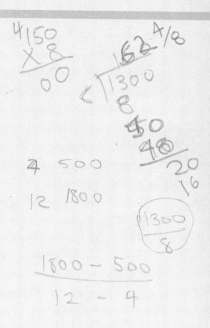

3.4 Predictions and Lines: $y = mx + b$

6. Nureen works in a department store during the summer vacation. Her weekly earnings are shown by the graph.

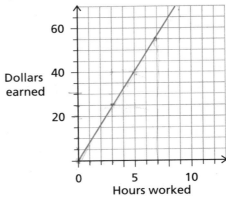

   (a) How much does she earn if no sales are made each week?
   (b) Write an equation to describe the relationship between hours worked and earnings.

7. The cost of printing the school newspaper is shown.

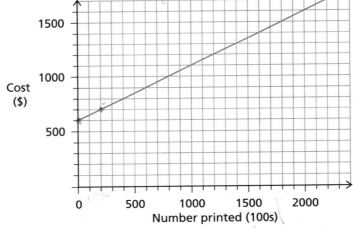

   (a) How much does it cost to print each copy of the school newspaper?
   (b) What is the cost to set up the printing equipment?
   (c) Write an equation to describe the relationship between number of copies printed and cost.

Think about...

Question 7

The data in Question 7 are discrete. The data points are connected on the graph. Why is this?

8. Fred was given equations that were to represent Internet providers.
   - State the hourly rate and the base monthly rate for each plan.
   - Sketch the graph of each equation.

   (a) $y = 2x + 5$ (b) $y = 4x - 1$ (c) $y = -5x + 7$
   (d) $y = 0.5x + 10$ (e) $y = 15 + x$ (f) $y = -2x + 8$

Think about...

Question 8

Which of the equations in Question 8 do you think would not represent Internet provider rates? Why?

9. Look at the equations that can represent Internet provider rates in Question 8.
   (a) Suppose you need about 13 h of Internet use per month. Which provider should you use? Why?
   (b) Suppose you can spend about $75 on an Internet provider each month. Which provider will you use? Why?

10. Write an equation for a line to represent each situation below. Then write a problem where this equation can be used. Use the concept of the Internet provider as the context for your problem.
    (a) slope = 3, y-intercept = 12.5   (b) slope = 0.6, y-intercept = 20
    (c) slope = 5, y-intercept = 0

11. The cost of renting a laptop computer is shown on the graph. Huran received a bill for $115. For how many days did he rent the computer?

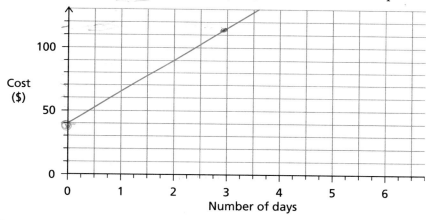

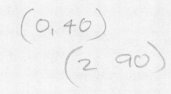

(0, 40)
(2, 90)

$y = mx + b$

+ 40

12. The labour charges for constructing a wheelchair ramp are shown in the graph.
    - What is the hourly cost?
    - If the total cost was $1872.50, how many hours of labour were involved in constructing the ramp?

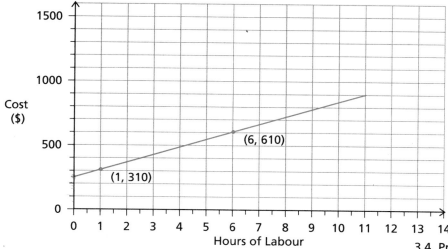

Writing and Using Equations

A student borrowed $490 to buy a stereo. She agreed to repay the loan at $35 per month. Write an equation to represent this line.

Method 1

- Construct a table of values for the amount owing after each month.
- Construct a graph to represent the table of values.
- Write the equation of the line. What do you notice about the slope?

Method 2

You already know that slope can be written as $\frac{y_2 - y_1}{x_2 - x_1}$.

Slope can also be represented by m. Thus, $m = \frac{y_2 - y_1}{x_2 - x_1}$.

Follow these steps to use $m = \frac{y_2 - y_1}{x_2 - x_1}$ and find the equation of the line. Summarize the process, along with an example, in your notebook.

- Find two points on the graph from Method 1, for example, (1, 455) and (2, 420).
- Find the slope of the line. $m = \frac{420 - 455}{2 - 1}$

  The slope is –35.

  This represents a $35 decrease each month.

Notice how the graph moves down from left to right. This represents a negative slope.

- The y-intercept is 490.
- The equation of the line is $y = -35x + 490$.

Focus Questions

13. The graph of $y = 5$ is a horizontal line.
    (a) Show the line on a grid.
    (b) What is the y-intercept? How do you know?
    (c) Explain why this graph has a slope of zero.

14. The graph of $x = 5$ is a vertical line.
    (a) Show the line on a grid.
    (b) What is the x-intercept? How do you know?
    (c) Explain why this graph has a slope that is **undefined**.

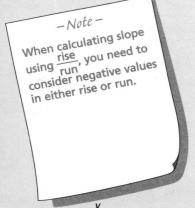

— Note —
When calculating slope using $\frac{rise}{run}$, you need to consider negative values in either rise or run.

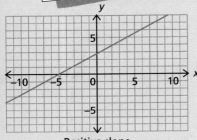

Positive slope

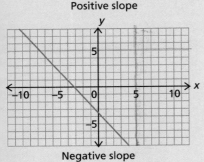

Negative slope

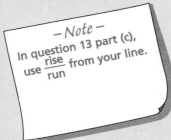

— Note —
In question 13 part (c), use $\frac{rise}{run}$ from your line.

undefined — a number divided by zero has no answer; division of a number by zero is undefined

Check Your Understanding

15. Describe the line given by each equation in as much detail as you can. How are the lines alike? How are they different?
    (a) $y = 4x - 5$ (b) $y = -5x + 2$ (c) $y = 2 + 7x$

16. Which would be steeper? Explain why.
    - a graph with a slope of $\frac{2}{3}$
    - a graph with a slope of $-\frac{5}{6}$

17. The food bank has 650 kg of potatoes. Each family that needs assistance receives a 5-kg bag.
    (a) Write an equation to represent the relationship.
    (b) How many families can receive potatoes?

18. In 1996, the St. John River overflowed its banks after reaching record heights. Water levels and rate of flow (in m³/s) are tracked by Environment Canada. Here is the rate of flow for seven weeks.

Week	1	2	3	4	5	6	7
Flow	900	1100	1175	1245	1310	1500	1700

    (a) Write an equation to represent the line of best fit.
    (b) What will be the rate of flow after 8.5 weeks? Is this answer reasonable? Give reasons for your answer.

    — Note —
    In Question 18, you could use graphing technology to plot these points and make your predictions. You will learn more about using this technology in Chapter 4.

19. A student keeps track of the time spent designing sets for the school drama production. After four days, she has spent 17 h. After 10 d, she has spent 35 h.
    (a) If she works the same number of hours each day, how many days are needed to complete the set if she estimates that a total of 100 h is needed?
    (b) Discuss what the intercept on the vertical axis might mean in this situation.
    (c) Suppose that instead of 100 h, she needs only 10 h. Show how you can use your graphing calculator to find the time needed.

    — Note —
    To make decisions, you may need to use your knowledge of how to solve equations. A graphing calculator can also help you.

20. Are we getting better as a nation at recycling? The amount of garbage recycled each day per person for various years is shown below.

Year	1960	1965	1970	1975	1980	1985	1990
Recycled (kg)	0.08	0.09	0.10	0.11	0.16	0.17	0.32

(a) Graph the data.

(b) Draw a line of best fit.

(c) Find the equation of the line.

(d) Predict the number of kilograms of goods that will be recycled each day in 2010.

Focus 1: Equations and the x-intercepts

The place where a graph crosses the *x*-axis can provide important information. You can find the exact point using your skills with solving equations.

Example 2

(a) The graph of $y = -5x + 2$ was graphed in Question 15. Find the *x*-intercept.

Solution
To find the *x*-intercept, substitute $y = 0$ into the equation of the line.

$$y = -5x + 2$$
$$0 = -5x + 2$$
$$0 - 2 = -5x + 2 - 2$$
$$-2 = -5x$$
$$\frac{2}{5} = x$$

Why does substituting 0 make sense?

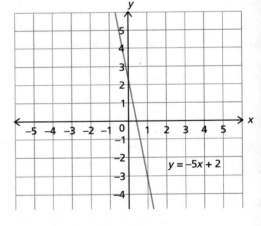

> *x*-intercept — the point where a graph crosses the *x*-axis; the point where $y = 0$

(b) Suppose a car traveled at an average speed of 50 km/h and uses 0.5 L of gas to travel 5 km. The car currently has 8 L of fuel. After how many kilometres will the car run out of fuel?

Solution
The graph of the situation is shown at the right. To find the total distance, you can find the *x*-intercept as that is the point where there is no fuel. From the graph, it appears the car can travel 80 km. Is there are another way to solve the problem?

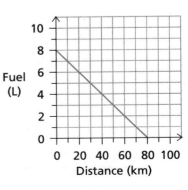

124 Chapter 3 *Patterns, Relations, Equations, and Predictions*

Check Your Understanding

21. For each line, find the equation, the *x*-intercept, and the *y*-intercept.
    (a) the line that passes through (2, 5) with a slope of 4
    (b) the line that passes through (–3, 1) with a slope of 0
    (c) the line with an *x*-intercept of 4 and a slope of $\frac{3}{4}$
    (d) the line with a *y*-intercept of 5 and a slope of $-\frac{1}{3}$
    (e) the line with an *x*-intercept of 2, *y*-intercept of –3

22. (a) Plot the points (–3, 5) and (1, –2).
    (b) Create an equation for the line through (–3, 5) and (1, –2).
    (c) Find the value of *x* when *y* is 12.

— Note —
In your notebook, summarize the process you used to find the equation of the line in Questions 21 and 22.

23. Adult tickets to a movie cost $7.50. Children's tickets cost $4.50. Suppose you have $42.50 to spend.
    (a) Write an equation that shows the number of tickets that can be purchased for *c* children and *a* adults. Graph the equation.
    (b) Suppose only children went to the movie. How can you use the equation to find the number of tickets that can be purchased? Find the number.
    (c) Repeat part (b) for only adults.

24. After running a marathon (42 km), Cynthia should consume 70 g of carbohydrates; no more and no less due to health concerns. One ounce of apple juice contains 3 g of carbohydrates. One pretzel contains 2.3 g of carbohydrates.
    (a) Write an equation to show the number of pretzels and the number of ounces of apple juice Cynthia can consume.
    (b) Suppose Cynthia only wanted something to drink. How many ounces of apple juice could she consume? How can you tell from the equation? From the graph?

25. As a furniture salesperson you earn commission and a base salary each week. For sales of $6000, you earn $630. For sales of $8000, you earn $790.
    (a) Find an equation to represent this situation.
    (b) From your equation, what is the base salary each week?
    (c) From your equation, what is the rate of commission?
    (d) Suppose you sold $12 000 worth of goods. How much would you earn? Describe how you found the amount.
    (e) Suppose you earned $950. What is value of the goods sold?

3.4 Predictions and Lines: $y = mx + b$

Investigation 6
What If ...?

Sandra's Internet provider charges $0.75 for each hour of use plus a monthly charge. She just received her first monthly bill for $30 for 20 h of use.

Purpose

Find the equation of a line from minimal information.

Procedure

A. Graph the situation.

B. How did you use the hourly charge of $0.75 per hour to draw the graph for the cost of Sandra's Internet service?

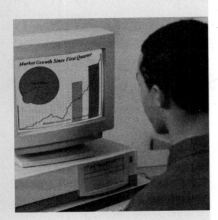

C. Suppose the costs associated with this Internet provider were for the first year only. After that, there is a choice:
- the hourly cost will increase to $1.50; or
- the monthly fee will be a fixed rate of $30.

How might each of these changes affect your graph?

Investigation Questions

26. Refer to the original problem.
- What information do you know that can help you to write the equation? How did you use this information?
- If you were to write the equation without a graph, how might you do so?

27. For Sandra's Internet provider, find the cost of using the Internet for:
(a) 15 h in one month during the first year.
(b) 35 h in one month during the first year.

Check Your Understanding

28. Craig rented 40 videos last year for a total cost of $170. The rental charge is $3 per video.

    (a) Sketch the graph that represents the relationship.
    (b) Without using the graph, find the equation of the line.
    (c) Verify your graph in (a) using this equation.
    (d) Discuss what the y-intercept in this graph represents.
    (e) Does this graph have an x-intercept? Does it make sense in this situation?

29. The cost of joining a skateboarding club is $10 plus $2 for every meeting attended. How would the graph for this relationship change if:

    (a) the membership cost were $20?
    (b) the cost to attend a meeting were $1.50?

30. Winnie has been investigating purchasing bread from three bakeries. They all charge for delivery.

Bakery	Delivery Charge	Cost per 100 Loaves Including Delivery	Cost for 200 Loaves Including Delivery	Cost for 300 Loaves Including Delivery
Baked Fresh	$5.00		$155	
Loaf Masters			$150	$220
Atlantic Bread		$73		

(a) Compare the Baked Fresh and Loaf Masters options. What is the cost per loaf for each bakery? What is the delivery charge for Loaf Masters?

(b) Winnie expects to order 350 loaves per week. Which option in (a) gives the best price?

(c) What additional information do you need to compare Atlantic Bread with Baked Fresh and Loaf Masters? How would you use this information?

(d) Create a value for Atlantic Bread so that it is the least expensive option for Winnie. Explain how you made your choice.

(e) Repeat part (d) to make Atlantic Bread the most expensive option.

3.4 Predictions and Lines: $y = mx + b$

Rearranging Equations and Formulas

If an equation is written in the form $y = mx + b$, it can be graphed quickly. If the equation is not written in the form $y = mx + b$, it is useful to rearrange it.

Example 3
Find the slope and y-intercept of $3x + 2y - 12 = 0$.

Solution

$$3x + 2y - 12 = 0$$
$$3x + 2y - 12 + 12 = 0 + 12 \qquad \text{Add 12 to both sides.}$$
$$3x + 2y = 12$$
$$3x - 3x + 2y = 12 - 3x \qquad \text{Subtract } 3x.$$
$$2y = 12 - 3x$$
$$\frac{2y}{2} = \frac{12}{2} - \frac{3x}{2} \qquad \text{Divide each term by 2 to isolate } y.$$
$$y = 6 - 1.5x$$

The y-intercept is 6. Interpret the new equation.
The slope is -1.5.

$y = -1.5x + 6$

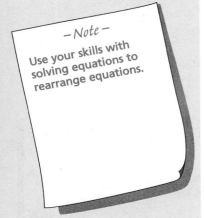

— Note —
Use your skills with solving equations to rearrange equations.

With the widespread use of spreadsheets and graphing calculators, it is important for you to be able to rearrange equations to solve for different variables. Spreadsheets and graphing calculators require you to enter equations in the form $y = mx + b$.

Suppose you know the area and the width of a room, and you need to find the length. You know that area = length × width. You need to rearrange the formula to find an expression for length.

Example 4
Solve $A = lw$ for l.

Solution

$$A = lw$$
$$\frac{A}{w} = \frac{lw}{w} \qquad \text{Divide } \textit{both sides} \text{ by } w \text{ to isolate the variable } l.$$
$$\frac{A}{w} = l \qquad \text{Simplify.}$$

128 Chapter 3 *Patterns, Relations, Equations, and Predictions*

Example 5
Solve $P = 2l + 2w$ for l.

Possible Solution

$P = 2l + 2w$

$P - 2w = 2l - 2w + 2w$ Subtract $2w$ from both sides
$P - 2w = 2l$ to isolate the term containing l.

$\dfrac{P - 2w}{2} = \dfrac{2l}{2}$ Divide both sides by 2 to isolate l.

$\dfrac{P - 2w}{2} = l$

Focus Question

31. Describe how rearranging a formula is similar to solving an equation.

Check Your Understanding

32. Velocity can be calculated from $v = u + at$ (initial velocity + acceleration × time). Suppose you know the velocity, initial velocity, and time. How would you find the acceleration?

33. How would you solve for work (w) using the formula $P = \dfrac{w}{t}$ (Power = work ÷ time)?

34. Solve each formula for the variable indicated.
 (a) $I = Prt$ for t
 (b) $P = 2l + 2w$ for l
 (c) $D = \dfrac{m}{v}$ for m
 (d) $A = \dfrac{1}{2}bh$ for b
 (e) $V = lwh$ for h
 (f) $V = \pi r^2 h$ for h

35. Which of the following has the steepest slope? The greatest y-intercept?
 (a) $3x - 4y - 12 = 0$
 (b) $2x + 5y - 30 = 0$
 (c) $5x - y - 9 = 0$
 (d) $7x + 3y - 21 = 0$
 (e) $4x - 5y - 12 = 0$
 (f) $x - 3y - 8 = 0$

3.5 More Patterns

Investigation 7
More Patterns and Graphs

Purpose
Explore a pattern and use it to create a model for tiling a floor.

Miguel tiled a floor using the pattern shown below. Each square tile is 0.5 m wide.

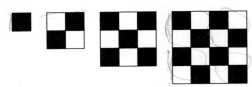

Procedure

A. Draw the next two pictures in the pattern.

B. Miguel created a table of values to summarize the pattern. Copy and complete the table of values. Describe any patterns that connect the number of tiles per side with the total area covered.

The Table
Describe how you found each total area.

Number of Tiles Per Side	Total Number of Tiles	Total Area Covered (m²)
1	1	0.25
2	4	1
3	?	2.25
4	?	?
5	?	?
6	?	?

C. Construct a graph to show the relationship between the number of tiles on one side and the total area covered. Identify the independent variable and the dependent variable. Explain how you made the decision.

D. How is this pattern like the patterns seen earlier in this chapter? How is it different?

E. Construct a graph to show the relationship between the number of tiles per side and the total number of tiles.

F. Compare your graphs in C and E. How are the graphs alike? How are they different?

G. Describe the pattern that connects the number of tiles per side to the total number of tiles. Express your pattern in words.

H. Describe the pattern that connects total area covered to the total number of tiles. Express your pattern in words.

Step F
Your graphs should not have an *x*-intercept or a *y*-intercept. Why not?

Investigation Questions

1. Suppose there were 12 tiles along one side.
   (a) How much area would be covered?
   (b) How did you find your answer in (a)?
   (c) How confident are you in your answer?
   (d) How might you check the accuracy of your answer?

2. (a) Predict the total area covered by 81 tiles.
   (b) How did you make your answer in (a)?
   (c) How confident are you in your answer?
   (d) How might you check the accuracy of your answer?

3. Miguel found that using a table of values to solve the problem required a large sequence of numbers. Do you think that a graph was a better way to try to solve Questions 1 and 2? Why?

4. Which method gives you the most precise answer?

K Writing and Using Equations

Miguel needs to find the equation for the total number of tiles he'll need from the number of tiles per side.

Refer to the table of values in Investigation 7.

- How are the number of tiles per side and total number of tiles related?
- Write an equation to connect the number of tiles per side and the total number of tiles. Verify your equation using the table of values.

He also wanted to find the equation for the total area covered from the number of tiles.

Miguel set up another equation to find the number of tiles to buy.

Area to be covered
= area of one tile × total number of tiles
= area of one tile × $\dfrac{\text{number of tiles}}{\text{along one side}}$ × $\dfrac{\text{number of tiles}}{\text{along adjacent side}}$

- Write an equation to show the area to be covered.

Miguel knew that the area to be covered was 150 m².
- Write an equation to find the number of tiles needed.

Focus Question

5. (a) Construct a graph that compares the total number of tiles used with the total area covered.
   (b) Write an equation that describes the relationship you graphed in (a).
   (c) Compare the graph in (a) to the graphs from Investigation 7. Why do they differ?

Check Your Understanding

6. You need to tile a square floor using square tiles that are 0.25 m wide. The area to be covered is 81 m².
   (a) Write a relationship to connect the total area covered and the number of tiles per side.
   (b) Graph the relationship.
   (c) How many tiles will you need?
   (d) Solve the equation that represents the situation. What does the solution represent?

7. The volume of a cylinder can be found using the formula $V = \pi r^2 h$, where r is the radius of the cylinder and h is its height.
   (a) Rearrange the formula to find r. You may need to refer to Focus J of Section 3.4.
   (b) If the height is 20 cm and the volume is 2000 cm³, find the radius.

Think about...

Question 7
Suppose you used a calculator to help you. Why does the π key give you a more accurate answer than solving the problem by hand?

Focus L

Connections

Julia's math teacher has given her the following table of values. Her task is to find an equation to describe the data.

x	0	1	2	3	4	5
y	3	0	−1	0	3	8

Julia decided to plot the data to look for a pattern.
- Plot Julia's points. Describe the shape of the graph.
- How is the shape of the graph like the one in Investigation 7? How is it different?
- Describe the shape of the graph using the word symmetry.

Elaine, another member of the group, said, "The y-value is 0 when x is 1 and again when x is 3. I remember that we called those x-intercepts. I wonder if they connect to the equation."

Samir replied, "I remember! When we found the x-intercepts, we were solving equations like $0 = 4x - 5$. Maybe the equations are for the relationship $0 = x - 1$ and $0 = x - 3$."

Julia said, "Using the distributive property to multiply $(x - 1)$ by $(x - 3)$ gives a result that has an x^2 term. How can we see if that works?"

Think about...

Julia's Comment
How do you think Julia might verify if she is correct? Does Julia's statement make sense?

Investigation 8

Representing the Situation

Purpose

Samir, Julia, and Elaine decided that $y = (x - 3)(x - 1)$ would give them the equation that represents the table of values. Determine if Samir, Julia, and Elaine are correct in saying that different equations can represent the same relationship.

– Note –
Samir, Julia, and Elaine found an expression that is in factored form. Will developed an expression that was in expanded form.

Procedure

A. Check the group's equation using the table of values. What does this suggest?

B. Then Will said, "I was looking at that pattern last night and I came up with $y = x^2 - 4x + 3$. I wonder if those two are the same."
- How might you check to see if Will is correct?
- Check this equation. What can you conclude?

3.5 More Patterns

C. Describe how you might verify your conclusions in A and B using graphs, then do so. Are your conclusions still valid?

D. Verify that $(x - 3)(x - 1) = x^2 - 4x + 3$ using algebra tiles. Record the representation in your notebook. What conclusions can you make?

Investigation Questions

Think about...

Question 8

Suppose you were to draw a graph of one part of this question. The lowest or highest point on one of these graphs is the vertex. Estimate the vertex on each graph. How does the vertex relate to the symmetry of the graph?

quadratic — an algebraic expression, equation, or relation that is of second degree

8. Which expressions are the same? Use tiles, tables of values, and/or graphs to help you.
   (a) $x^2 + 3x + 2$ and $(x + 1)(x + 2)$
   (b) $x^2 - 4x - 5$ and $(x - 5)(x - 1)$
   (c) $x^2 + 7x + 6$ and $(x + 6)(x + 1)$
   (d) $x^2 + 5x + 4$ and $(x + 1)(x + 4)$
   (e) $x^2 - 7x - 6$ and $(x - 6)(x - 1)$

9. Write the x-intercept and the corresponding **quadratic** relation for each part of Question 8. How do these relate to the factors of the expression?

FOCUS M Algebra Tiles and Expressions

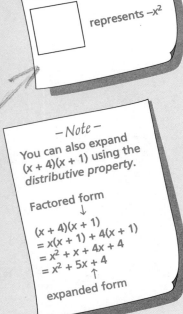

— Note —

represents x^2

represents $-x^2$

In Investigation 8, the group saw that two different expressions might give the same rectangle when modeled with algebra tiles. They wanted to know if they could create a factored expression from an expanded expression or an expanded expression from one that is already factored. They decided to use algebra tiles to explore these expressions.

$(x + 4)(x + 1)$ $(2x + 1)(x + 3)$ $3x^2 + x - 4$

Example 1
Write $(x + 4)(x + 1)$ in expanded form.

Solution
Step 1 Model the situation as shown.

Step 2 Count the algebra tiles, then write the new expression.

$(x + 4)(x + 1) = x^2 + x + 4x + 4$
$\qquad\qquad\qquad = x^2 + 5x + 4$

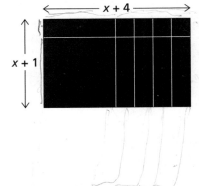

— Note —
You can also expand $(x + 4)(x + 1)$ using the distributive property.

Factored form
↓
$(x + 4)(x + 1)$
$= x(x + 1) + 4(x + 1)$
$= x^2 + x + 4x + 4$
$= x^2 + 5x + 4$
↑
expanded form

134   Chapter 3 *Patterns, Relations, Equations, and Predictions*

Example 2
Write $(2x + 1)(x + 3)$ in expanded form.

Solution
$(2x + 1)(x + 3) = 2x^2 + 7x + 3$

- How do the algebra tiles demonstrate this solution?
- How might you find the polynomial for this expression using the distributive property?
- Which method do you prefer?

Example 3
Write $3x^2 + x - 4$ as a product of two factors.

Solution
Step 1 Identify the algebra tiles you need: three x^2-tiles, one x-tile, and four negative-unit tiles.

Step 2 Use the tiles to try to create a rectangle. Is it possible? What might you do to make it possible?

Step 3 Use the tiles you now have from step 2. Create a rectangle. How do the tiles show that $3x^2 + x - 4$ can be written as $(3x + 4)(x - 1)$? Be specific in your description.

– Note –
In step 2, remember that whatever tiles you add must give you a net result of zero.

Writing $3x^2 + x - 4$ as $(3x + 4)(x - 1)$ means that you are writing it in factored form. The factors are $(3x + 4)$ and $(x - 1)$.

Check Your Understanding

10. Expand.
    (a) $(x + 5)(x + 2)$ (b) $(2x + 3)(x + 4)$ (c) $(x - 2)(x - 4)$
    (d) $(3x - 2)(x + 5)$ (e) $(x - 4)(x + 4)$ (f) $x(x + 5)$

11. Find the area of each rectangle.
    (a) length = $2a + 3$, width = $a + 2$ (b) length = $x + 1$, width = $2x + 1$
    (c) length = $5 - w$, width = $w + 2$ (d) length = $y - 2$, width = $y - 3$

12. (i) Use algebra tiles to model each pair of expressions.
    (ii) Use the model to decide which pairs are equivalent.
    (iii) How might you prove that two expressions are equivalent without using tiles?

    (a) $2x + 6$ $2(x + 3)$
    (b) $5(2x - 3)$ $10x + 15$
    (c) $3x(2x - 1)$ $5x^2 - 3x$
    (d) $(x + 4)(x + 3)$ $x^2 + 7x + 12$
    (e) $x^2 - 6x + 9$ $(x - 3)^2$
    (f) $(x - 3)(x + 3)$ $x^2 - 9$
    (g) $(2x + 1)(x - 5)$ $2x^2 - 4x - 5$

Think about...
Question 13
Why do you think factoring expressions (a), (b), and (c) can be called common factoring?

Think about...
Question 14
Two of the factors of 10 are 5 and 2. Thus, $\frac{10}{5} = 2$. How does this property help you simplify these expressions?

Think about...
Question 16
This is called factoring the **difference of squares**. Describe how it is done. What patterns can help you factor the expression without using tiles?

13. Factor.
    (a) $12x + 6$
    (b) $3x - 9$
    (c) $x^2 + 5x$
    (d) $x^2 + 4x + 4$
    (e) $x^2 - 2x + 1$
    (f) $x^2 - 6x + 9$

14. Simplify each expression.
    (a) $\frac{12x + 6}{6}$
    (b) $\frac{3x - 9}{3}$
    (c) $\frac{4x - 16}{4}$
    (d) $\frac{x^2 + 4x}{x}$
    (e) $\frac{x^2 - 6x}{x}$
    (f) $\frac{x^2 + 5x}{x}$

15. Show that $(2x + 3)(x + 1)$ and $2x^2 + 5x + 3$ are equivalent.

16. You are going to factor $x^2 - 1$.

     $(x+1)(x-1)$

    (a) What is the first step?
    (b) Carl created this rectangle. He added one positive *x*-tile and one negative *x*-tile. Why do you think he did this?
    (c) Write an expression to show the factors of $x^2 - 1$.
    (d) Factor each of the following expressions.
        (i) $x^2 - 9$
        (ii) $x^2 - 16$
        (iii) $x^2 - 25$
        (iv) $25x^2 - 100$
        (v) $x^2 - 64$
        (vi) $3x^2 - 12$

Investigation 9
Patterns

Purpose
Look for patterns that connect the expanded and factored forms of an expression.

Procedure
Look at the following.

expanded form		factored form
$x^2 + 3x + 2$	=	$(x + 2)(x + 1)$
$x^2 - 3x + 2$	=	$(x - 2)(x - 1)$
$x^2 + 4x + 3$	=	$(x + 3)(x + 1)$
$x^2 + 2x + 1$	=	$(x + 1)(x + 1)$

A. What patterns do you see that can be used to get the factored form from the expanded form of an expression?

B. Copy the chart into your notebooks for each expression in this Investigation. For each polynomial:
   (a) Construct and record an algebra-tile model.
   (b) Graph the equation formed by setting each expression equal to y.
   (c) Complete the corresponding chart entries.

Expression	x^2 Tiles	Unit Tiles	x-Tiles	Grouping of x-tiles	x-intercept	y-intercept	Factors
$x^2 + 3x + 2$	1	2	3			0 2	
$x^2 - 3x + 2$	1	2	-3			0 2	

C. Describe the connection between the grouping of the x-tiles and the x-intercepts of the corresponding graphs.
D. Describe the connection between the factors of the expressions and the grouping of the x-tiles.
E. Describe how the y-intercept is connected to the original expression.
F. Write a method for factoring a polynomial of the type shown above.
G. Repeat Steps A to F for the following.
$2x^2 + 9x + 4 = (2x + 1)(x + 4)$
$4x^2 - 3x - 1 = (x - 1)(4x + 1)$
$6x^2 - x - 2 = (2x + 1)(3x - 2)$
$10x^2 - 19x + 6 = (5x - 2)(2x - 3)$

Then describe how the factors of a polynomial help you find the x-intercepts of the graph of the polynomial.

FOCUS N: Factoring Expressions of the Form $ax^2 + bx + c$

As Elaine pointed out, the x-intercepts are the points on a graph for which $y = 0$. These can be found from a quadratic equation in the form $ax^2 + bx + c = y$, when $y = 0$. By factoring the expression on the left side of the equation $ax^2 + bx + c = 0$, we can determine the x-intercepts.

You can use the pattern from Investigation 9 to factor expressions algebraically.

Example 4
Factor $3x^2 + 11x + 6$.

Solution

3.5 More Patterns 137

You can also write the following:

$3x^2 + 11x + 6 = 3x^2 + 9x + 2x + 6$
$= 3x(x + 3) + 2(x + 3)$
$= (3x + 2)(x + 3)$

Explain what happened in each step.

Example 5
Factor $4x^2 - 4x - 3$.

Solution
Repeat the process used in Example 4.

$4x^2 - 4x - 3 = 4x^2 - 6x + 2x - 3$
$= 2x(2x - 3) + 1(2x - 3)$
$= (2x - 3)(2x + 1)$

Remember to explain what happens in each statement you write and each calculation you make.

Check Your Understanding

17. Factor each expression.
    (a) $2x^2 + 3x - 9$
    (b) $3x^2 + 8x + 4$
    (c) $2x^2 - 11x + 12$
    (d) $2x^2 + 5x + 2$
    (e) $2x^2 + 7x + 6$
    (f) $2x^2 - 7x + 3$

18. Solve each equation using the patterns from Investigation 9. Check your work with algebra tiles. How do the algebra tiles help you to verify the patterns?
    (a) $x^2 + 6x + 5 = 0$
    (b) $x^2 + 8x = 0$
    (c) $x^2 + 7x + 6 = 0$
    (d) $x^2 - 1 = 0$
    (e) $x^2 + 8x + 15 = 0$
    (f) $3x^2 + 8x + 4 = 0$
    (g) $25x^2 - 100 = 0$
    (h) $2x^2 + x - 6 = 0$

19. Solve each equation. Check your factoring with a graph. What do you notice about the solutions to the equations and the x-intercepts on the graphs?
    (a) $x^2 - 5x - 14 = 0$
    (b) $x^2 + 3x = 10$
    (c) $x^2 - 8x + 16 = 0$
    (d) $x^2 - 9 = 0$
    (e) $-5 + 7x = 2x^2$
    (f) $-5x + 4x^2 - 6 = 0$

20. Factor.
    (a) $t^2 + 11t$
    (b) $20 + x^2 - 9x$
    (c) $p^2 - 10p - 24$
    (d) $3x^2 - 9x$
    (e) $10d + 25$
    (f) $-100 + q^2$
    (g) $x^2 + 10x + 25$
    (h) $x^2 - 16$
    (i) $2x^2 - x - 10$
    (j) $28x + 3x^2 + 9$

21. (a) The area of a rectangle is $w^2 + 10w + 25$. Find the length of each side. Model this expression using tiles.
    (b) Repeat for $4x^2 + 12x + 9$.

22. (a) Write the area equation for a rectangle that is four times as long as it is wide. Define any variables you use.
    (b) If the area is 16 m², find the length and width of the rectangle.

23. Use your knowledge of algebra tiles to simplify each expression. Record the algebraic steps to simplify these expressions using your tiles as a guide.
    (a) $\dfrac{x^2 + 7x + 12}{x + 4}$
    (b) $\dfrac{x^2 + 5x - 14}{x + 7}$
    (c) $\dfrac{2x^2 + 11x + 12}{x + 4}$
    (d) $\dfrac{3x^2 + 5x - 2}{3x - 1}$
    (e) $\dfrac{9x^2 + 6x + 1}{3x + 1}$
    (f) $\dfrac{x^2 - 9}{x - 3}$

24. If a quadratic equation like $x^2 + 3x + 1 = 0$ will not factor, can it be solved? Explain.

25. (a) Show how you would use graphs to solve the equation $x^2 - 2x - 6 = 2$.
    (b) Explain why you can't solve this equation as it now looks using the method from Investigation 9.
    (c) Fred says that this equation can be changed by subtracting 2 from each side to get $x^2 - 2x - 8 = 0$, and that this is an equivalent equation. Is Fred correct? Why?
    (d) Fred says that this polynomial can be factored as
      $x^2 - 2x - 8 = (x - 4)(x + 2)$.
      Show that Fred's factoring is correct.
    (e) Fred claims that, after factoring, he can now see the solution to the original problem right away. Explain how he can get the answer so quickly.

Think about...

Question 21

The expression is called a perfect square trinomial. Look at the factors. Why do you think this is so?

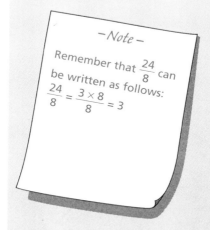

— Note —
Remember that $\dfrac{24}{8}$ can be written as follows:
$\dfrac{24}{8} = \dfrac{3 \times 8}{8} = 3$

CHALLENGE yourself

The graph of $y = x^2 + 6x + 10$ is shown below. How do you think this graph helps you decide whether the expression $x^2 + 6x + 10$ can be factored?

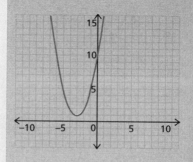

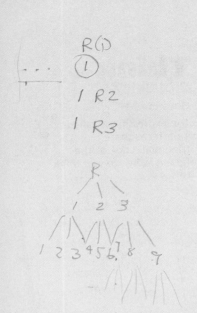

26. David plans to get into shape by doing pushups. It's been a while since he exercised regularly, so he plans to get into it gradually.

On the first day he does 10 pushups. The next day he does 13; the third day 16; the fourth day 19, and so on.

(a) How long it will take him to reach 100 pushups in one day?

(b) If he starts on October 1 and sticks to his plan until the end of the month, how many pushups will he be doing on October 31?

(c) Describe how you can tell quickly that this relationship is linear.

27. How is solving $x - 4 = 0$ like solving $x^2 - 7x + 12 = 0$? How is it different? Use the method discussed in this section to help with your explanation.

Chapter Project

Designing a Dog Pen

Seana must think about the area of her rectangular dog pen so that her dog, Brownie, will have as much space to run as possible.

(a) Calculate the area for your pen designs. Plot a graph of width (independent variable) versus area.

(b) Which model gives the maximum area? How confident are you that this is the maximum area? Explain.

(c) Write an equation for the graph in (a).

(d) The vet suggests an area of 28 m² to help keep Brownie healthy. Can Seana build this pen? Explain.

(e) Find the least amount of fencing to build a pen of area 45 m². Would a pen this big be necessary?

(f) If Seana uses the back wall of her home as one side of the pen, find the maximum area for a perimeter of 24 m.

(g) How can a graph of width versus area for a pen help Seana to find possible values for the length and width?

3.6 Other Patterns

Samir asked Julia and Elaine, "I wonder if there are any other patterns with curved graphs besides those in Section 3.5. For example, I saw a commercial once that said, "I told two friends and she told two friends, and so on, and so on." Does this form a curved pattern?"

Investigation 10
Other Patterns

Purpose
Explore Samir's question above.

Procedure
The director of the school musical changed the rehearsal schedule at the last minute. There was not enough time for one person to call each of the cast, crew, and band. The director decided to phone two cast members. These two members would, in turn, phone two other members, who, in turn, would phone two other members, and so on.

A. Construct a tree diagram to show the calls. Describe the pattern.

B. Plot a graph to show the round number on the horizontal axis and the number of calls made on the vertical axis.

C. Describe any patterns you see in the graph.

D. In which round will 64 people be called? How do you know?

E. Write an equation to model the pattern. How can you use your equation to find in which round 256 people would be called? Find the solution.

Investigation Questions

1. (a) Does the graph have an *x*-intercept? A *y*-intercept? Does this make sense in the context of the problem?
   (b) Write the domain and range. How do the domain and range fit the context of the problem?

2. (a) What is the independent variable?
   (b) What is the dependent variable?
   (c) Are your choices for independent and dependent variable reasonable in the context of the problem? Give reasons.

---

tree diagram — a pictorial way of representing or modeling combinations of things

For example, in Step A, the first two rounds are as follows:

round 1: two people called director

round 2: four people called director

person 1   person 2
  / \   / \
 3   4   5   6

Think about...

Step E
How confident are you in your solution? How might you verify your solution?

3. (a) Suppose each cast member made three phone calls each instead of two. Describe how the graph might change.

   (b) Suppose each cast member made one phone call instead of two. Describe how the graph might change. Have you seen a graph like this previously? Where?

Check Your Understanding

4. (a) Estimate the student population of your school.

   (b) Gossip can spread very quickly in a school. Suppose one person in your school started a rumour and told four other people. These four each told four others who had not yet heard the rumour, and so on. How many rounds would it take until everyone in your school heard the rumour?

   (c) Suppose, in part (b), only two people were told each time. Does this mean that twice as many rounds are needed? Why?

   (d) What do you notice about the number of rounds in (b) and (c)? Does this suggest a pattern?

Exponential Relations and Equations

In Investigation 10, you drew the graph of an **exponential relation**. In the previous section, you drew graphs of quadratic equations.

exponential relation — a relation in the form $y = a^x$, where a is any positive number

A. Julia asked Samir and Elaine how the graphs are alike and how they are different. Help Samir and Elaine compare the graphs. Think about:
   - intercepts;
   - shape of the graph; and
   - domain and range, and so on.

B. Samir drew the graph of the exponential equation $y = 3^x$ and constructed a table of values for the equation. How can this graph help with the explanation in step A?

x	y
−2	0.111
−1	0.333
0	1
1	3
2	9
3	27

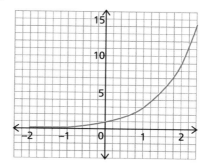

Think about...

The Table of Values

How did Samir use his skills with exponents to complete this table of values? Use your calculator in your explanation.

C. Julia then asked, "By drawing graphs we have found solutions to equations. This is probably no different. What is the value of x when y = 81?"

$$81 = 3^x$$
Since $81 = 3^4$,
then $3^4 = 3^x$.
Thus, $x = 4$.

D. Finally, Julia had a brainstorm: "I bet our skills with equations can be combined with Step C to solve exponential equations." Julia showed her reasoning.

$$3^x + 1 = 244$$
$$3^x + 1 - 1 = 244 - 1$$
$$3^x = 243$$
Since $243 = 3^5$,
then $3^5 = 3^x$.
Thus, $x = 5$.

Think about...

Step D

Is Julia's reasoning correct? Give reasons for each statement she wrote to show that her reasoning is either correct or incorrect.

Focus Questions

5. If you know only the equation, how can you decide whether the graph will show an exponential equation or a quadratic equation?

6. (a) Sketch a graph of $y = 2^x$ and $y = x^2$.
   (b) Compare the graphs as Samir, Julia, and Elaine would.

Did You Know?

This book was made from signatures of 16 pages. A signature is a large piece of paper that is folded to create 16 pages.

Check Your Understanding

CHALLENGE yourself

What do you think is the largest signature that can be made to create a book? Why?

Fold a piece of paper to create a signature of 16 pages. Number the pages so that, when the page is folded again, a book of 16 consecutive pages is formed. Describe any patterns in your numbering.

CHALLENGE yourself

Someone in your class has picked a page number in this book.

(a) Develop a plan that will allow you to ask questions and identify the page number in as few steps as possible.

(b) How did you use your understanding of exponential equations to develop your plan?

(c) Try your plan. Does it need to be modified?

(d) Elaine claimed that the maximum number of questions needed to identify the page number is eight. Is Elaine correct? Why?

---

7. For a biology experiment, the number of bacteria in a sample doubles every hour. If there are 200 bacteria at the beginning of the experiment, how many will there be in
   (a) 1 h?   (b) 2 h?   (c) 4 h?   (d) 6 h?

8. A type of yeast cell doubles every 20 min. If there are 350 yeast cells in a batch of bread dough originally, how many will there be in
   (a) 20 min?   (b) 1 h?   (c) 2 h?

9. Fold a piece of paper in half as many times as you can.
   (a) What is the greatest number of times you can fold the paper? Were the last two folds difficult?
   (b) Write an equation that shows the number of rectangles created by each fold.
   (c) Write the domain and range of the equation.
   (d) Suppose a friend asked you to determine the number of rectangles created for 12 folds. Your friend needs to know this for a science-fair project. What answer might you give him?

10. In an old game show called "The $64 000 Question," contestants received money for correct answers. The first correct answer was worth $125. The second was worth $250. For each correct answer, the money doubled. How many correct answers were necessary to win the grand prize?

11. Solve each equation. Use your skills with mental math.
    (a) $3^x = 27$ (b) $5^x = 25$ (c) $2^x = 8$
    (d) $10^x = 1$ (e) $4^x = 16$ (f) $2^x = 64$
    (g) $2^x + 5 = 133$ (h) $3^x - 2 = 7$

12. For each equation, identify the type of graph as linear, exponential, or quadratic. Then solve the corresponding equation.
    (a) $y = x^2 + 5$; solve for $y = 174$
    (b) $y = 2x + 5$; solve for $y = -22$
    (c) $y = 2^x$; solve for $y = 128$
    (d) $y = 4^x + 2$; solve for $y = 66$
    (e) $y = 6 - 3x$; solve for $y = 9$
    (f) $y = x^2 - 6x$; solve for $y = -9$

PUTTING IT TOGETHER

CASE STUDY 1

Every September, a Hot Air Balloon Festival is held in Sussex, New Brunswick. These colourful balloons are able to fly because the air inside the balloons is heated to approximately 100°C. The hot air in the balloons is less dense than the air surrounding them, so the balloons rise. One balloon rises at a rate of 36 m/min.

(a) Create a table of values to compare the height of the balloon and the time. Describe any patterns you see.

(b) Sketch a graph of height versus time. Would the graph continue indefinitely along this path? Why or why not?

(c) Write a linear equation in the slope y-intercept form to represent the graph. What do the slope and y-intercept represent?

(d) What other factors would have an effect on the height of the balloon?

(e) How long will it take the balloon to reach a height of 1.7 km? Explain your method and show all your work.

(f) The air temperature decreases by about 3°C for each kilometre you rise above the ground. If the air temperature at ground level is 18°C, when should the pilot put on a winter jacket? Explain your answer.

CASE STUDY 2

Create a math problem using facts about your community. The solution of the problem must require an equation.

To find the necessary community information, you will have to contact and visit a community business, organization or visit a web site such as:

- a local business
- a municipal government office
- a food bank
- a sports team
- a second hand clothing store
- and so on

Putting It Together   145

Before you contact the organization, think of some questions you would like to ask about what they do. For example, if you are visiting a food bank, you may want to find out how many people use it each day, week, or month.

At least one fact you learn must contain numerical data that will be used in solving your problem.

Each assignment must include all of the following:

- a paragraph with at least five facts about the business or organization;
- the number you have chosen to be the answer to the problem, with a brief explanation of what the problem is about;
- the process you used to form your equation;
- the problem; and
- the written solution.

EXTENSION

The Confederation Bridge, linking Prince Edward Island to New Brunswick, opened May 31, 1997. There are 44 piers supporting it. Assume the first pier closest to the shore is 78 m high. Each pier is 1 m taller than the previous one up to the centre pier. The piers going down on the other side decrease in size in the same pattern as those going up. The bridge is 12.9 km long.

(a) Set up a table of values to show the relationship between the pier number and the height of that pier.

(b) Graph the relationship.

(c) What is the slope of the graph for the first half of the bridge? What does it represent? What would the slope be for the second half of the bridge? Explain.

(d) What does the y-intercept of the graph represent? Explain. Discuss whether or not there should be an x-intercept. If so, what would it mean?

(e) The speed limit on the bridge is normally 80 km/h. How long will it take to cross the bridge? Show all your work.

(f) Write a short paragraph describing the importance of the bridge to Prince Edward Island and New Brunswick.

REVIEW

Key Terms

	page
common factoring	136
continuous data	97
difference of squares	136
discrete data	97
distributive property	113
domain	97
equation	102
exponential relation	142
factoring	135
factors	135
identity	113
intersection point	111
linear	100
modeling	96
perfect square	139
quadratic	134
range	97
sequence	96
slope	117
slope y-intercept form	117
tree diagram	141
undefined	122
verify	108
x-intercept	124
y-intercept	117

You Will Be Expected To

- define variables and create expressions, relations, and equations, including linear and quadratic.
- write expressions, relations, and equations from patterns, tables of values, verbal descriptions, and problems.
- create problem situations from graphs and equations.
- simplify or expand expressions, relations, and equations.
- factor.
- use graphing to solve equations, and to find x-intercepts and y-intercepts.
- plot and recognize the graphs of linear and quadratic relations, including distinguishing between discrete and continuous values.
- solve linear and quadratic equations.
- use factoring as a tool in solving equations and graphing.
- work with equations of non-linear relations.

Summary of Key Concepts

3.1 Describing Patterns

Patterns can often be found by **modeling** a situation. Models can be scale models, a table of values, a graph, or even an algebraic relation.

Example 1
Three shapes are constructed using blocks below.

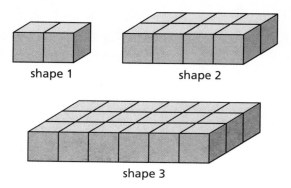

shape 1   shape 2

shape 3

(a) Record the length, width, perimeter, and area for the top of each shape in a table of values.
(b) Model the next three shapes to continue your tables of values.
(c) Construct a graph to show each relationship:
length versus width   perimeter versus area
length versus area   length versus perimeter
(d) Which graphs show a linear pattern? a non-linear pattern?
(e) Find the area when the perimeter is 54 cm. How did you find your answer?

Solution
(a) and (b)

Length (cm)	Width (cm)	Perimeter (cm)	Area (cm²)
2	1	6	2
4	2	12	8
6	3	18	18
8	4	24	32
10	5	30	50
12	6	36	72

(c) and (d)
Length vs. Width (linear)

Perimeter vs. Area (non-linear)

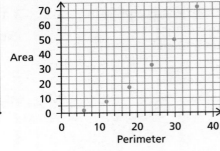

Length vs. Area (non-linear)

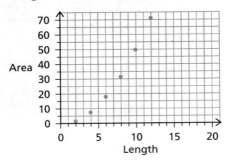

Length vs. Perimeter (linear)

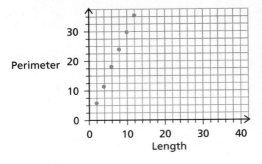

(e) The area is 162 cm².

3.2 Solving Problems by Solving Equations

Solving equations allows you to get more accurate answers and make more accurate predictions about future events.

Example 2

Use the pattern in Example 1.
(a) Write an equation to show the relationship between length and perimeter.
(b) Write a relation to show the relationship between length and area.
(c) Find the length when the perimeter is 231 cm.

Solution

(a) From the graph:
$$\text{slope} = \frac{\text{rise}}{\text{run}}$$
$$= \frac{3}{1}$$
$$= 3$$

The y-intercept is 0.

Thus, the equation can be written as $y = 3x$ where y is the perimeter and x is the length.

(b) The graph is non-linear. From the table of values, it appears that
$$\text{area} = \text{length} \times \frac{1}{2}\text{length.}$$
$$= \frac{1}{2}\text{length}^2$$

(c) $y = 3x$
$$231 = 3x$$
$$\frac{231}{3} = \frac{3x}{3}$$
$$77 = x$$

3.3 Decision Making and Patterns

To find the solution to an equation like $3x = 5x - 3$, follow these steps:
- Graph $y = 3x$.
- Graph $y = 5x - 3$ on the same grid.
- Estimate the point of intersection.

3.4 Predictions and Lines: $y = mx + b$

You can write the equation of a line even when only partial information about a situation is known.

Example 3

Gene is a sales representative with a large department store. He earns a base salary plus commission. His earnings in the last two weeks were:

Sales: $10 000   Earnings: $400
Sales: $15 000   Earnings: $500

(a) Find Gene's commission rate and base salary.

(b) Write an equation for the line.

(c) Gene earned $475 last week. What were his total sales?

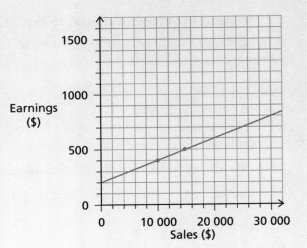

Solution

Gene plotted the points as shown on the graph.

(a) His commission rate can be found from the slope of the line.

$$\text{slope} = \frac{\text{rise}}{\text{run}}$$

$$= \frac{100}{5000}$$

$$= 0.02 \qquad \text{Gene's commission rate is 2\%.}$$

His base salary can be found by extending the line from the graph to estimate the y-intercept. Gene's base salary is $200, per week.

(b) The equation of the line is $y = 0.02x + 200$, in which y is earnings and x is total sales.

(c)
$$y = 0.02x + 200$$
$$475 = 0.02x + 200$$
$$475 - 200 = 0.02x + 200 - 200$$
$$275 = 0.02x$$
$$\frac{275}{0.02} = \frac{0.02x}{0.02}$$
$$13\,750 = x \qquad \text{Gene's total sales last week were \$13 750.}$$

3.5 More Patterns

Sometimes, the relation that models a situation is not linear.

Example 4

(a) Show how to use a graph to solve $x^2 - 2x - 6 = 2$.

(b) Fred says that this equation can be changed by subtracting 2 from each side to get $x^2 - 2x - 8 = 0$, and that this is an equivalent equation. Is Fred correct? Why?

(c) Fred decided that this equation can be factored as $x^2 - 2x - 8 = (x - 4)(x + 2)$. Show that Fred's factoring is correct.

(d) Fred claims that he can now see the solution to the original problem right away. Explain how he could get the answer so quickly.

Solution

(a) Look at where $y = x^2 - 2x - 6$ crosses $y = 2$.

(b) Yes. Subtracting 2 from both sides maintains the balance.

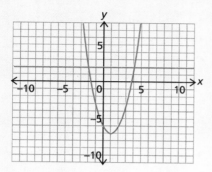

(c)

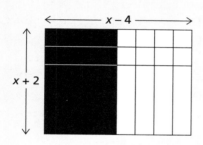

(d) To equal zero, at least one of the factors equals zero.
$x - 4 = 0$, so $x = 4$
$x + 2 = 0$, so $x = -2$

3.6 Other Patterns

Example 5

In 1980, the Peregrine Falcon population had been reduced to about 75 by the use of pesticides. Since then, with the restriction on pesticide use, the population has increased by about 12% per year. Estimate the population in this year.

Year	Number of Falcons
1980	75
1981	75 + 12% of 75 = 84
1982	84 + 12% of 84 = 94
1983	94 + 12% of 94 = 105

Solution

To help estimate the population, draw a graph.

A. Use this table of values to draw a graph.

B. Extend your graph.
  - What pattern do you see?
  - Predict the number of falcons there will be this year.

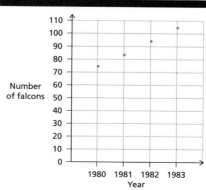

PRACTICE

3.1 Describing Patterns

1. Continue the following pattern. Copy and complete the table of values.

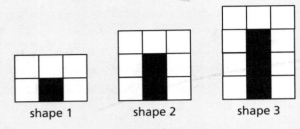

shape 1   shape 2   shape 3

Shape Number	Number of Black Squares	Number of White Squares	Total Number of Squares
1			
2			
3			
4			
5			
6			
7			

(a) Which shape in this pattern has a total of 120 squares?

(b) Find the number of white squares in the shape that has 24 black squares.

(c) Find the number of black squares in the shape that has 93 white squares.

(d) Is there a shape that has exactly 249 white squares? Explain how you arrived at your decision.

3.2 Solving Problems by Solving Equations

2. As manager of the school store, you have been asked to prepare a report for the student council that predicts the sales of T-shirts for the next month. You know that:

- you sold six T-shirts the first day you were open; and
- each day your sales increased by five T-shirts.

(a) Show the T-shirt sales relationship using a table of values and a graph.

(b) On which day did you sell 76 T-shirts?

(c) Which shows your increase in sales more effectively: the graph or the table of values? Explain.

3. Anne and her friends take a taxi to the Crash Test Dummies concert, 12 km from their school. The cost of a taxi is $2.00 to start and $0.60 for every minute.

(a) Find the total cost of the 23-min ride to the concert. Assume a part of a minute is charged a part of the $0.60.

(b) Show the relationship between cost and time.

   (i) What method did you decide to use?

   (ii) What other methods could you have used?

(c) Anne graphed this relationship. Should she join the points of the graph? Explain.

(d) The cost of returning home was $17.00. How long was the trip?

4. Another cab company would have charged Anne and her friends $3.50 to start and 50¢ for every minute. Assume a part of a minute is charged a part of the $0.50.

(a) Graph this relationship on the same axes as the graph in Question 3. Compare the two graphs.

(b) Explain what the difference in slope of the two graphs tells you.

(c) Explain the meaning of the point where the graphs cross. This is called the intersection point.

(d) What values are allowed for the trip times?

(e) How does your answer in (d) affect what quadrants you used for the graph?

3.3 Decision Making and Patterns

5. Solve each equation using algebra tiles.
   (i) Draw the algebra tile diagram for each step.
   (ii) Write the algebraic version of each step.
   (a) $5x - 3 = 2x - 9$
   (b) $4x + 8 = 2x + 4$
   (c) $-3x + 1 = 2x - 9$
   (d) $4 - 2x = x + 1$
   (e) $4 - 3x = x + 1$

6. Solve each equation using the method of your choice. Before solving, predict which equation has the greatest value.
   (a) $3(x + 7) = 11(x - 1)$
   (b) $\frac{2}{3}(x - 1) = \frac{1}{3}(x + 4)$
   (c) $-0.8(2x + 5) = x - 6.6$
   (d) $11x - 13 + 4x = 10x - 5 + 7x$
   (e) $\frac{x - 2}{3} = \frac{28 - 2x}{6}$
   (f) $\frac{x + 1}{5} = \frac{3x - 2}{14}$

3.4 Predictions and Lines: y = mx + b

7. The student council sells T-shirts for $12 each and sweat shirts for $25 each. They sold 20 more T-shirts than sweat shirts. Their sales last year totaled $1720. How many sweatshirts did they sell?
   (a) Two pieces of information in this problem can be used to create equations. What are they?
   (b) What equation(s) should you graph in order to solve the problem?
   (c) Find the solution and check it using your equation(s).
   (d) Check to make sure your solution makes sense for the situation in the problem.

8. There are many pairs of numbers whose average always equals 2.
   (a) Set up a table of values with at least 12 pairs that average 2. Some sample values are provided to get you started.

First Number	Second Number	Average
3	1	$\frac{(3 + 1)}{2} = 2$
0	4	$\frac{(0 + 4)}{2} = 2$
−1	5	$\frac{(-1 + 5)}{2} = 2$

   (b) Write the equation that gives the value of the second number in terms of the first.
   (c) Graph your results. Use the first number for x and the second for y.
   (d) Show how the equation would change if the numbers had to average 10.
   (e) Show how the graph would change if the numbers had to average 10.
   (f) How would the equation and the graph change if the average had to be 7?
   (g) How would the graph and the equation change if the average had to be −3?

3.5 More Patterns

9. (a) Write expressions to represent the perimeter and the area of the following rectangle.
   (b) Find the values of the perimeter and the area if $w = 5$ cm.

10. Factor each expression using algebra tiles. Draw a labeled diagram for your answer.
    (a) $x^2 + 6x + 5$
    (b) $3x^2 + 8x + 4$
    (c) $x^2 - 3x - 10$
    (d) $x^2 - 1$
    (e) $2x^2 - x - 6$

11. Factor each expression using a graph. Use a graphing calculator if available.
    (a) $x^2 - 5x - 14$
    (b) $4x^2 + 6x - 18$
    (c) $x^2 - 8x + 16$
    (d) $4x^2 - 9$

12. Factor each expression algebraically.
    (a) $t^2 + 11t + 18$
    (b) $x^2 - 9x + 20$
    (c) $p^2 + 10p - 24$
    (d) $x^2 - 9x - 22$
    (e) $d^2 + 10d + 25$
    (f) $q^2 - 100$
    (g) $x^2 + 4x + 4$
    (h) $12x^2 - 80$

13. Solve each equation.
    (a) $x^2 - 2x - 15 = 0$
    (b) $x^2 - 4 = 0$
    (c) $5x - 4 = 2x + 8$
    (d) $x^2 + 5x = 14$
    (e) $6x - 5 + 2x = 4x + 7$
    (f) $x^2 - 12 = 4x$

3.6 Other Patterns

14. Solve each equation.
    (a) $5^x - 4 = 121$
    (b) $3^x + 20 = 101$
    (c) $5 + 2^x = 37$

15. A newborn bass typically has mass of about 0.3 g. In a hatchery, the fish increases in mass by about 10% per week. The fish can be put into a stocked lake when it reaches a mass of about 0.5 kg. After how many weeks can it be put into the stocked lake?

16. (a) Write the area equation for a rectangle that is three times as long as it is wide.
    (b) If the area is 4 cm², find the length and width of the rectangle.

Chapter Four
Modeling Functional Relationships

If you study hard for your next test, what do you think the effect will be on your mark? If you smoke, what do you think the effect will be on your health? If you get a post-secondary education, how do you think this will affect your earning power?

In Chapter 1, you studied many cause-and-effect relationships. In this chapter, you will investigate cause-and-effect relationships further to find out how you might confidently predict the effect when you know the cause.

After successfully completing this chapter, you will be expected to:

1. Analyze graphs and tables to derive specific information, and write stories from graphical displays.

2. Describe a specific cause-and-effect relationship called a function, and use the function to make predictions about future events.

3. Predict a function that can represent graphed data.

4. Use graphing technology to determine functions that can represent specific data, and make predictions about future events using this function.

5. Describe your confidence level about your predictions from the models of relations you build.

4.1 Tables, Graphs, and Connections

Have you ever heard the saying, "a picture is worth a thousand words?" Graphs are a picture of data. In this section, you will investigate how to create and interpret the "story" graphs are telling.

Investigation 1
Create a Graph

Purpose
To create your own graph from collected data

Procedure

A. Select someone to be the walker. Have the walker stand in front of the teacher's desk and choose a path around the room with 10 stopping points.

B. Time how long it takes the walker to walk from the teacher's desk to the first point and each of the other points. Measure and record the distance walked, the time it takes to walk between the points, and the length of time the walker stops at each point.

— Note —
In step B, the speed at which the walker moves can change.

C. Copy and complete the table below.

Elapsed time (s)						
Distance walked (m)						
Time stopped (s)						

D. Graph the data you collected.

Investigation Questions

Think about...

Step D
Explain how the changes in the walker's speed will appear on the graph.

1. Refer to your graph.
   (a) Identify the points that correspond to the times when the person was moving the fastest.
   (b) Identify the points that correspond to the times when the person was stopped.

2. Suppose you were to ask someone else to follow the route the walker took, using only your graph as a guide.

   (a) How accurately do you think that person could follow the route used by the walker? Explain.

   (b) How accurately could that person copy the speed and timing of the walker? Explain.

Creating a Broken-Line Graph

Allan landed at Halifax International Airport (Robert L. Stanfield) and took the hotel courtesy bus to his hotel. The graph below shows the distance of the bus from the airport over time. It also shows that the speed of the bus varied as it traveled along its route. These data are shown with a broken-line graph

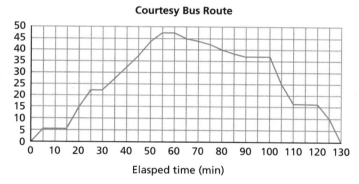

- The graph is completed by connecting the points with line segments. What do the points represent on the graph?
- Describe the resulting appearance of the graph.
- What does the highest point on the graph represent?

broken-line graph – a graph formed by joining data points with line segments

Think about...

The Graph

Can a graph have two maximum points?

What do you think the lowest point on a broken-line graph represents?

— Note —
Speed represents the change of distance over time and velocity is speed with a direction. A negative velocity indicates a movement in the opposite direction.

Focus Questions

3. You can find the slope of a line between any two times on the graph of the bus trip.

   (a) What does the slope represent?

   (b) How can you use slope to describe the movement of the bus?

4. If you look at the first half of the courtesy-bus graph, most of the slopes are positive. The slopes for the second half are mostly negative. What does this tell you about the trip?

Check Your Understanding

5. Lana kept track of the distance she was from her home each day while on a sightseeing vacation.

   (a) Construct a graph of the data.

Day	1	2	3	4	5	6	7	8	9	10	11
Distance (km)	200	400	700	750	750	600	500	400	250	200	0

   (b) Write a description of Lana's trip.

6. After finishing school, Jake started a company to promote concerts featuring local bands. He secured a line of credit with a bank to help finance his business. The graph shows the balance owing on his line of credit after each of the first 12 years of operation.

   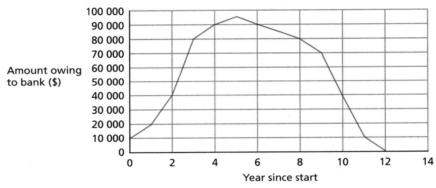

   (a) Briefly interpret the meaning of:

   (i) the maximum point on the graph;

   (ii) the slopes of the segments of the graph; and

   (iii) the difference in interpretation of a negative and a positive slope.

   (b) Use the graph to write a brief financial summary of the first 13 years of Jake's company.

Trend Lines

Joaquin has accepted an after school job as a sales associate in a clothing store. He needs to prove to his employers that he has a good math sense so that they will be confident in his ability to work the cash register.

As a test, Joaquin's employers give him a graph of his potential earnings based on sales and ask him to interpret it. Help Joaquin to interpret the graph by answering these questions.
- What is the value of the slope? What does the slope tell you?
- What is the value of the y-intercept? What does the y-intercept tell you?

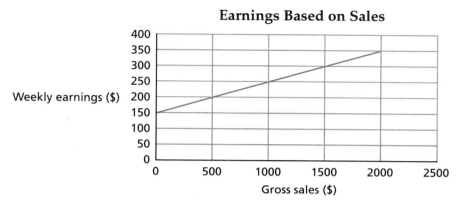

– Note –
In Chapter 3, you worked with linear graphs. These are similar to broken-line graphs except that they have just one segment. Just like a broken-line graph, the slope of a linear graph represents a rate of change. Both types of graphs usually have a y-intercept that represents an initial condition or starting value.

The line on the graph can be called a trend line. As you saw in Section 1.6, a line of best fit is often used as a trend line. Why would it be given that name?

To celebrate his new job, Joaquin's father took him to the drag races. Joaquin has a great interest in car racing, particularly drag races, and always analyzes the distances and times of his favourite drivers. While at the races, he recorded distance versus time for Dan Pastorini, his favourite driver, and analyzed the data on the graph shown below. He noticed that not all trends are linear.

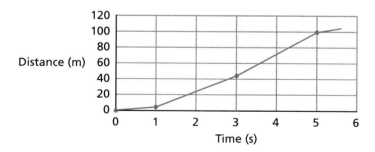

- Describe how Dan's speed changes over time.
- Estimate the slopes from 0 s to 1 s, from 2 s to 3 s, and so on. What do they tell you?

4.1 Tables, Graphs, and Connections 159

Check Your Understanding

7. The graph below shows the distance versus time for a runner in a short race. Use your knowledge about slopes to write a description of the race for this runner.

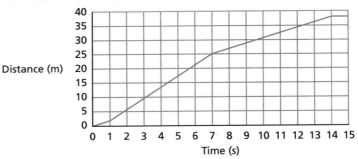

8. The graph of speed versus time of a race car as it goes around the track in the diagram is shown below.

   (a) What do the lowest points at the "troughs" of the curve represent?

   (b) What do the horizontal parts at the "humps" of the curve represent?

   (c) Describe one lap around the track and how the graph represents your description.

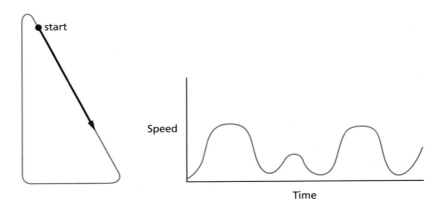

9. (a) Draw a graph of speed versus time as a car makes one circuit of each track. The arrow shows the direction the car is moving.

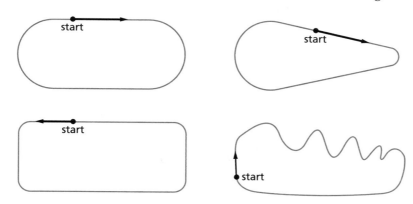

(b) Choose a partner and exchange graphs for one of the tracks. See if you can associate the graph with the appropriate track.

10. The graph shows the temperature in °C of a room over a 24-h period.

(a) Why is the graph a saw-tooth shape and not a horizontal line?

(b) In each "tooth," the two sides are not the same shape. Explain why.

(c) The last two "teeth" are not level with the rest. Give reasons why this could happen.

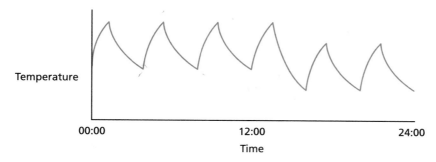

11. Jacinta ran to buy a newspaper from the corner store. The graph below shows her distance from home at various times.

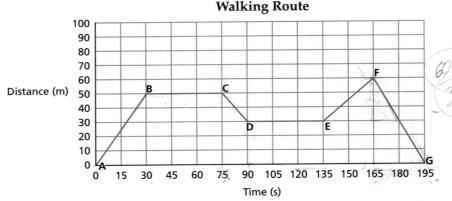

(a) What was her fastest speed? How long did she run at this speed?

(b) For how long was she stopped? How can you tell from the graph?

12. The graph shows the population of Arctic hares in an area that was once a mining community. With reforestation, other animals such as foxes and owls began moving into the area.

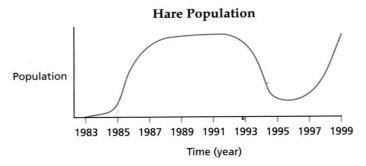

(a) When did the hare population reach its maximum number?

(b) When did the hare population reach its minimum number?

(c) Write a story that explains the behaviour of the graph.

13. (a) Draw a graph that shows the height of the water versus time as you fill a bathtub. On your graph, also show what happens when you get into the tub, and what happens when you pull the plug.

(b) Check your prediction at home.

14. (a) Select a glass and sketch its shape. Sketch a graph that shows the height of the glass of water versus the mass of water used. Check your sketch by actually carrying out an experiment. Compare your graphs. How accurate was your graphed prediction?

(b) Repeat part (a) using another glass shape. Compare your final graphs. How are they alike? How are they different?

4.2 Relations and Functions

Freda has joined four of her friends in a fitness group during her gym class. Their improvement in running capacity is being tested every Friday. In addition to regular training, they also jog for 20 minutes in preparation for the Friday test. For five weeks, Freda's group has recorded how far each runner can go in a seven-minute run during gym class.

The gym teacher was interested in the relationship between the number of weeks of jogging and running capacity. As capacity is estimated by the distance run in seven minutes, the distances run in seven minutes were recorded each Friday for five weeks.

Distance Run in Seven Minutes

Week	Freda	Vanita	Liz	Tracy	Ruth
1	0.5 km	1.2 km	0.8 km	0.75 km	1.0 km
2	0.6 km	1.1 km	1.0 km	0.9 km	1.0 km
3	0.7 km	1.3 km	1.2 km	1.0 km	1.2 km
4	0.8 km	1.5 km	1.4 km	1.2 km	1.3 km
5	0.9 km	1.7 km	1.5 km	1.4 km	1.5 km

— Note —
These data will be used for Investigation 2.

Investigation 2
Relations in General

Purpose
To investigate the relationship between jogging and running capacity

Procedure

A. Graph all of the results shown by Freda's group for each week on one coordinate grid.

B. Explain whether there appears to be a relationship between weeks spent jogging and the distance run in seven minutes.

— Note —
It is important that you place all graphs in step A on one coordinate grid. It can help to use a different symbol or colour for each individual member.

— Note —
The domain includes all of the values for the independent variable and the range includes all of the values for the dependent variable.

Investigation Questions

1. Predict how far each member of the group will run in week 6. Explain how you made your prediction.

2. Look at the graphs for each individual member of the group done in step A. Do the graphs for the individual members support your conclusions in step B? Give reasons for your answer.

3. From your relationship predict how far Vanita will run in seven minutes during week 10. Discuss with others how reasonable your prediction is. Use the domain and range in your discussion.

Functions

In Investigation 2, you could not predict how far a group could run. You could only predict how far the individuals could run because each week had many pieces of data. You can easily display the information for any one participant. For example, the data for Freda can be shown in different ways.

As a Graph

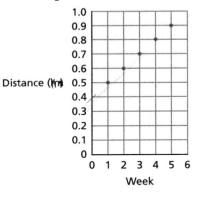

As a Table of Values

Week	Distance (km)
1	0.5
2	0.6
3	0.7
4	0.8
5	0.9

As an Equation

Freda's data seem to fall in a straight line. From your work in Chapter 3, you know that you can write an equation for this line as shown below.

$$\text{slope} = \frac{0.6 - 0.5}{2 - 1} = 0.1$$

From the graph, if there were a y-intercept, it appears that it would be 0.4.

The equation can be written as $y = 0.1x + 0.4$ where y is the distance run in kilometres and x is the week number.

As you saw in Freda's graph, each week number is paired with exactly one distance that she ran. As there seems to be a pattern in this pairing of numbers, it seems reasonable to say, "the distance Freda can run in seven minutes is a function of the number of weeks that she has been jogging."

function – a function is a relation for which every value of the independent variable is paired with one and only one value of the dependent variable. The statement that "y is a function of x" means that y is dependent on x. Therefore, x is the independent variable and y is the dependent variable.

The graph at the right shows the results for a member of another group. Notice that it also suggests a functional relationship. Describe the relationship. How is it different from Freda's?

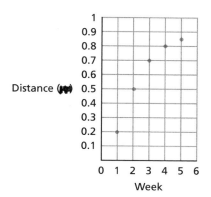

Think about...

Data for Freda's Entire Group

Suppose you used the mean distance run for each week. Could you now make predictions about the distance run in week 6? Draw the graph for the entire group using mean distance for each week. Would it be reasonable to say that the mean distance the entire group can run is a function of the number of weeks the girls have been jogging? Give reasons for your answer.

The graph at the right shows the same relationship between distance run and number of weeks of jogging for another group. The individuals who produced these times are not identified. Explain why the relationship is *not a function*.

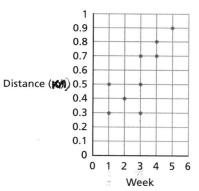

D Using Models Expressed as Functions

A function can always describe something that is real. Mathematicians often use the notation $f(x)$, read as "f of x" or "f at x", when naming functions. Instead of writing $y = x^2$, you can write $f(x) = x^2$. This says that y is a function of x. To rewrite an equation using $f(x)$ notation is called "rewriting the equation using function notation." To do so, replace y with $f(x)$.

$$y = x + 2$$

$$f(x) = x + 2$$

Example 1

Two restaurant waiters disagreed about whether candles always burn at the same rate. One of the waiters thought that candles burned faster as they got smaller. On a quiet evening, they made the measurements shown.

Time (min)	30	60	90	120
Height of candle (mm)	120	114	108	102

Predict the height of the candle after 4 h.

Solution

Here is a graph of the results. It shows the candle height y as a function of the time of burning x.

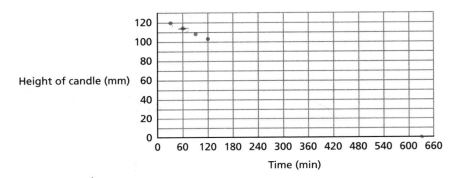

Based on your work in Chapter 3, explain why the equation of the above relationship is $y = -0.2x + 126$.

Since the height of the candle is a function of time, you can write the linear function as $f(x) = -0.2x + 126$, where x is time in minutes.

The height of the candle after 4 h can be represented by $f(240)$. To find $f(240)$, substitute $x = 240$ into the function.

$$f(x) = -0.2x + 126$$
$$f(240) = -0.2(240) + 126$$
$$= 78$$

The height of the candle would be 78 mm after 4 h.

Think about...

The Graph in Example 1

Suppose that the first waiter is correct. What will be the shape of the graph? How will it be different if the second waiter is correct?

Focus Question

4. The following measurements were made to confirm whether the candle burned faster as it got smaller or it burned at the same rate throughout.

Time (min)	420	450	480	510	540	570
Height of candle (mm)	42	36	30	24	18	12

Which waiter was more likely to be correct?

Example 2
After how long will the candle burn out completely?

Solution
By reading the graph, you can estimate that the candle's height is 0 mm after burning for about 630 min. Or you can use the fact that the candle's height is a function of time.

$$H(t) = -0.2t + 126$$
$$0 = -0.2t + 126$$
$$0.2t = 126 \qquad \text{Add } 0.2t \text{ to both sides.}$$
$$t = 630 \qquad \text{Divide both sides by 0.2.}$$

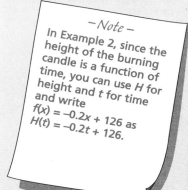

– Note –
In Example 2, since the height of the burning candle is a function of time, you can use H for height and t for time and write $f(x) = -0.2x + 126$ as $H(t) = -0.2t + 126$.

Focus Questions

5. Lana looked at the graph for the height of the burning candle. How can she decide quickly if the graph represents a function? Be specific in your description.

6. Refer to the graph of $f(x) = x^2$ shown below. Does your description in Question 5 also apply to this graph? If so, explain why. If not, modify your description so that it does. Does $f(x) = x^2$ represent a function?

Graph of $f(x) = x^2$

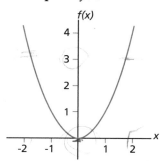

7. Explain why it is appropriate to say that in the solution of Example 2 you found the "zero of the function."

Think about...

Variables

The tables in Questions 8 to 10 show measurements of quantities that may be related. For each, identify the independent and dependent variables. What convention has been applied to the order that the variables appear in the tables?

Check Your Understanding

8. Sketch the graph represented by each table of values. Do the graphs represent functions? Give reasons. Are they linear or non-linear?

   (a) The altitude of a parachute jumper is related to the time that the parachutist has been falling.

Time (s)	10	15	20	25	30
Height (m)	7010.00	6397.50	5540.00	4437.50	3090.00

   (b) A cylinder's height is twice its radius. Its volume is related to its radius.

Radius (cm)	2	4	6	8	10
Volume (cm³)	50.24	401.92	1356.48	3215.36	6280.00

   (c) The number of plants in a square garden is related to the length of the garden.

Length (m)	3	5	7	9	11
Number of plants	72	200	392	648	968

– Note –
When a tree is cut down, the stump shows many rings. The number of rings shown is the age of the tree. The diagram below shows a tree that is six years old.

9. Brian and Craig examined their Christmas tree and wondered if the diameter of the tree trunk might be a function of its age in years. They went to a Christmas-tree farm and looked at the tree stumps for all of the trees that had been cut down. They used the growth rings to make these measurements.

Age (years)	5	7	9	11	13	15	17	19
Diameter (cm)	7	11	15	19	23	27	31	35

   Write a mathematical expression to show that tree trunk diameter is a function of age.

Think about...

Question 10

Why should it not surprise you that this graph shows a linear function?

10. Andrew and Jennifer wondered if the amount of fuel pumped at a gas pump was a function of the time that the pump was running. While Jennifer filled the tank, Andrew twice took note of the amount of fuel pumped. He started recording after Jennifer had begun pumping the gas.

Elapsed Time (s)	10	20
Volume pumped (L)	22	44

(a) Write a mathematical relation to represent the volume pumped as a function of time.

(b) Graph the relation. Why does the graph represent a function?

11. The perimeter P of this triangle is a function of the side length s. Write an equation for the function.

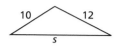

12. Describe a situation between two variables so that one variable is a function of another. Then describe another situation in which a relationship between two variables is not a function.

13. Find the value of the function in each of the following.

(a) $d(t) = 20t + 3$ Find $d(3)$.
(b) $f(x) = 2x^2 + 3x$ Find $f(-1)$ and $f(2)$.
(c) $g(x) = 5x^2 + 3x - 2$ Find $g(-2)$ and $g(1)$.
(d) $T(t) = 240 \times 0.7^t$ Find $T(2)$.
(e) $A(t) = 1000 \times 1.07^t$ Find $A(10)$.
(f) $P(n) = \dfrac{64}{n}$ Find $P(4)$.

14. Suppose you knew that a relation was a function given by $f(x) = 2x + 8$. Describe how you could find the value of the independent variable x if you knew that the value of the function is 16.

15. Find the missing value of the independent variable in each of the following linear functions.

(a) $v(T) = 332 + 0.6T$ $v(?) = 340$
(b) $L(T) = 4400 + 0.05T$ $L(?) = 4410$
(c) $n(s) = 0.75s + 3.5$ $n(?) = 12.5$
(d) $g(x) = -0.32x + 8.7$ $g(?) = 1.02$
(e) $p(L) = \dfrac{4}{5}L + 9$ $p(?) = 37$

16. For each set of data shown in the tables,
- identify the independent and dependent variables;
- plot a graph for the data; and
- decide whether each graph represents a function.

(a) Revenue of a Concert

Number of people admitted	0	20	20	40	40	60	60
Revenue of concert ($)	0	100	200	300	400	500	600

Think about...

Question 11

All triangles have the property that the sum of the lengths of any two sides is greater than the length of the third side. What does this tell you about the domain and range of the perimeter function?

4.2 Relations and Functions

x = ind.
y = dep.

(b) **Ice Cream Sales**

Highest daily temperature (°C)	20	20	20	20	20	20	20
Number of ice creams sold	0	5	10	15	20	25	30

I / d

(c) **Oven Temperature with Time**

Time oven is on (s)	0	10	20	30	40	50
Temperature (°C)	20	110	180	230	260	270

I / d

(d) **Skid Mark and Car Velocity**

Length of skid mark (m)	0	20	20	40	40	60	60
Velocity of car (km/h)	0	30	−30	45	−45	57	−57

d / I

(e) **Cooling of Coffee**

Time elapsed (s)	0	30	60	90	120	150	180
Temperature of coffee (°C)	95	75	61	50	42	36	32

I / d

17. How can you change one of the non-functional relationships in Question 16 to a function that you can use for making predictions? Use the definitions of independent and dependent variables and their relation to each other for your discussion.

FOCUS E — Graphs That Represent Functions

— Note —
For the relation describing Freda's group at the top of the next page to be a function, every value of the week number can correspond to only one value of distance run in seven minutes.

To make predictions from data, you need to know that one variable is a function of another. Graphing the results is a quick way to find out if a relation is a function. If there is more than one value of the dependent variable (*y*-value) for each value of the independent variable (*x*-value), then the relation between the two variables is not a function and the data cannot be used to make a prediction.

170   Chapter 4 *Modeling Functional Relationships*

For example, the graph of data for Freda's entire group (shown below) does not represent a function. However, the graph of data for Freda herself does represent a function.

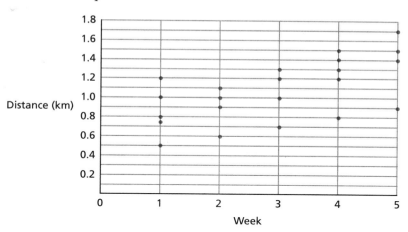

Think about...

Freda's Graph

Visualize a vertical line sweeping across the graph of Freda's data. What is the maximum number of points that the line will intersect at any one time?

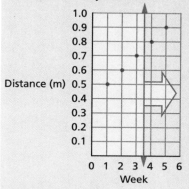

Notice that the time value for week 1 is paired with several values of distance run. You can show this as a mapping (as shown in the margin). Notice that there are several arrows coming from the value of week 1.

An easy way to decide whether the graph of a relation represents a function is to place a vertical line on the coordinate grid and sweep it across the graph. If the line intersects more than one point at a time anywhere on the graph, then the relation is not a function. The vertical line will always intersect the graph of a function at only one point. This is called the **vertical-line test**.

Both of the graphs below are functions.

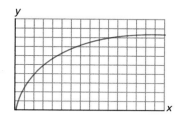 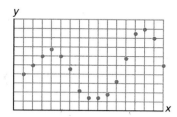

Neither of these graphs is a function.

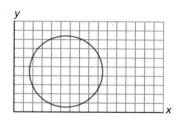 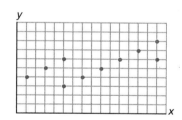

Think about...

The Mapping Diagram

What would the mapping diagram look like if the relation was a function?

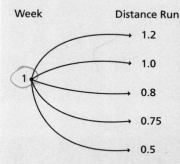

4.2 Relations and Functions   171

Focus Questions

18. Summarize in a sentence or two the vertical line test.

19. Give reasons why the vertical line test is a quick way for you to decide whether the graph of a relation represents a function.

Check Your Understanding

20. Explain why the vertical line test shows that this relation is a function.

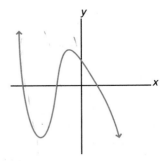

21. Explain why the vertical line test shows that this relation is not a function.

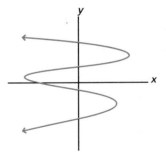

22. Use the vertical line test to decide which of these relations are functions. For those that are not functions, explain why.

(a)    (b)

172   Chapter 4 *Modeling Functional Relationships*

(c)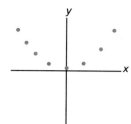

(d)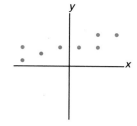

23. Use the vertical line test to decide which of these relations are functions. For those that are not functions, explain why.

(a)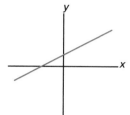

(b)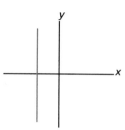

(c)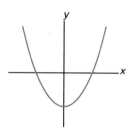

(d)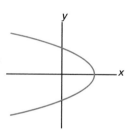

(e)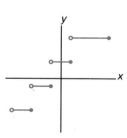

(f)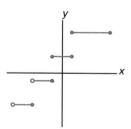

4.3 Equipping Your Function Toolkit

Technology

In this and all other investigations in this section, you will need to construct tables of values, plot the points, and look for patterns on the graphs. If technology is available, use it to handle the tedious parts of your work.

If you can recognize the graphs of the basic functions like $f(x) = x$ or $f(x) = x^2$, you can often use these basic shapes to sketch the graphs of more complex functions.

Investigation 3

Changes in Functions, Part 1

Purpose

To develop an understanding of the relationship between the basic graph of a quadratic function and the algebraic change in the equation of the graph of the function

Procedure

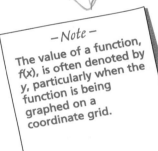

— Note —
The value of a function, $f(x)$, is often denoted by y, particularly when the function is being graphed on a coordinate grid.

A. Construct a table of values for each of the following quadratic functions. Use x-values between -4 and 4 such that x is an integer.

   (i) $f(x) = x^2$ (ii) $-f(x) = x^2$

B. Compare the values of the function for each x-value in the two tables. What do you notice?

C. Graph the two functions on the same coordinate grid using $y = f(x)$ for all real values of x. Describe the relationship between the two functions.

Investigation Questions

1. For each function below, make a table of values using x between -4 and 4. Then graph the functions on the same coordinate grid for all real values of x.

   A: $y = x^2 + 1$ B: $-y = x^2 + 1$

2. Are the two functions in Question 1 related in the same manner as you observed in step C?

3. (a) For two functions to be related in this manner, what appears to be the only difference between the equations that represent the functions?

   (b) Complete this sentence. When you see a $-y$ in the equation then the image is a ▬▬▬▬ in the x-axis.

174   Chapter 4 *Modeling Functional Relationships*

4. In step A, you explored the effects of changing y to $-y$ in $y = x^2$. You can describe the relationship between the two equations using the **mapping notation** $(x,y) \to (x,-y)$. Describe the relationship between the two functions in Question 1 using mapping notation.

> **mapping notation** – a notation that describes how a graph and its image are related

> *– Note –*
> In Question 4, $(x,y) \to (x,-y)$ is read as "x maps onto x, and y maps onto $-y$." This means that all x values stay the same but all y values are multiplied by -1.

Check Your Understanding

5. Decide whether each of the following non-linear graphs is the graph of a quadratic function. Give reasons to support your decision.

(a)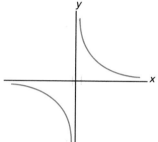

(b)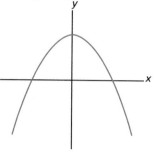

(c)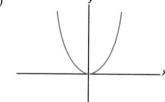

(d)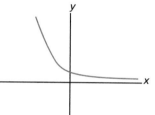

(e)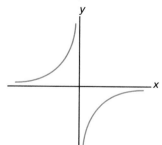

(f)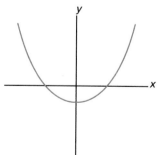

(g)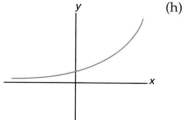

(h)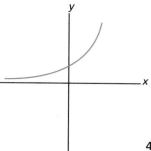

4.3 Equipping Your Function Toolkit

Investigation 3
Changes in Functions, Part 2

Purpose
To develop an understanding of the relationship between the basic graph of a quadratic function and the algebraic change in the equation of the graph of the function

Procedure

A. Construct a table of values for each of the following functions. Use x-values between -4 and 4 such that x is an integer.

   (i) $y = x^2$ (ii) $2y = x^2$ (iii) $\frac{1}{3}y = x^2$

B. Refer to the tables of values that you constructed for part (i) in step A. How do the y-values of $2y = x^2$ and $\frac{1}{3}y = x^2$ compare to the y-values of $y = x^2$?

C. Graph each function. How do the graphs compare?

D. Complete these sentences.

   "When I see $3y$ in the equation, then I know that there is a vertical stretch of ▉."

   "The vertical stretch factor is the ▉ of the coefficient of y."

Investigation Questions

6. Describe the relationship between the functions in step A using words and mapping notation.

7. Using your observations from this investigation, describe using words and mapping notation the transformation of $y = x^2$ that resulted in each of the following functions. Which function has the widest graph? Explain.

   (a) $3y = x^2$ (b) $5y = x^2$ (c) $\frac{1}{3}y = x^2$

8. Predict the shape and then sketch the graph given by each of the following mappings of $y = x^2$.

   (a) $(x,y) \to (x, \frac{1}{4}y)$ (b) $(x,y) \to (x, \frac{2}{5}y)$

Check Your Understanding

9. Khaled has been looking for a job. He had interviews with four different companies and they each had a different bonus plan for their staff which could be modeled by the equations below. In each case, y is the amount of bonus in thousand dollars and x is the number of years of service.

   (a) $2\frac{1}{2}y = x^2$ (b) $y = \frac{1}{4}x^2$

   (c) $10y = x^2$ (d) $y = x^2$

   - Match each model to an appropriate graph.
   - Which graph represents the best bonus for the same number of years worked?
   - Why would you have to ignore the left half of each graph?

   (i)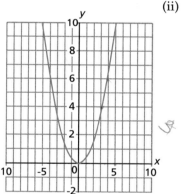

   (ii)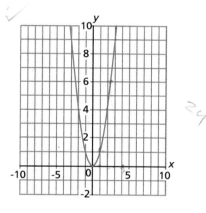

   (iii)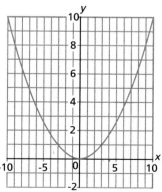

   (iv)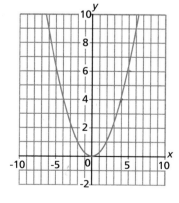

10. The graph of $y = x^2$ and an image graph of it are shown below. Write a mapping to relate the two graphs. Then write the equation of the image graph.

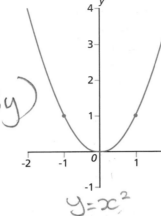

$y = x^2$

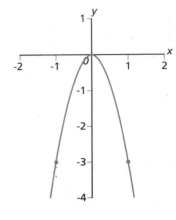

$(x, y) \rightarrow (x, -3y)$

$-\frac{1}{3}y = x^2$

— Note —
The vertical stretch factor is the multiplicative inverse of the coefficient of y.

11. Describe the transformation of $y = x^2$ that has resulted in each of the following image equations in words and using a mapping rule.

(a) $2y = x^2$

(b) $-\frac{1}{3}y = x^2$

(c) $-\frac{3}{5}y = x^2$

(d) $-5y = x^2$

12. Find the equation of the image of $y = x^2$ under each of the following mappings.

(a) $(x,y) \rightarrow (x, 5y)$

(b) $(x,y) \rightarrow (x, -4y)$

(c) $(x,y) \rightarrow (x, -\frac{1}{3}y)$

13. (a) Explain how you could use the graph of $y = x^2$ to help you to sketch the graph of $-y = x^2$.

(b) Draw the graph of $\frac{1}{3}y = x^2$.

(c) Explain how you could use the graph of $\frac{1}{3}y = x^2$ to help you to sketch the graph of $-\frac{1}{3}y = x^2$.

Investigation 3

Changes in Functions, Part 3

Purpose

To develop an understanding of the relationship between the basic graph of a quadratic function and the algebraic change in the equation of the graph of the function

Procedure

A. Construct a table of values for each of the following functions. Use x-values between -4 and 4 such that x is an integer.

(i) $y = x^2$ (ii) $y + 2 = x^2$ (iii) $y - 3 = x^2$

B. Refer to the tables of values that you constructed for part (i) in step A. How do the y-values for $y + 2 = x^2$ and $y - 3 = x^2$ compare to the y-values of $y = x^2$?

C. Graph each function. How do the graphs compare?

D. Complete these sentences.

"When I see $y + 2$ in the equation, then I know that there is a vertical translation of ■ units ■."

"When I see $y - 3$ in the equation, then I know that there is a vertical translation of ■ units ■."

"The vertical translation is the ■ of the number being added to the y."

Investigation Questions

14. Describe the transformational relationship between each pair of equations in words and using mapping notation.

(a) $3(y + 2) = x^2$ and $y = x^2$
(b) $y - 5 = x^2$ and $y = x^2$
(c) $-(y + 3) = x^2$ and $-y = x^2$
(d) $-2(y + 1) = x^2$ and $y = x^2$

15. Which equation describes each of the graphs in (a), (b), and (c)? How do you know?

A: $-2(y - 1) = x^2$ B: $y - 3 = x^2$ C: $-(y - 2) = x^2$

(a)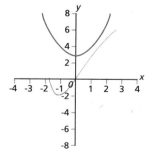

(b)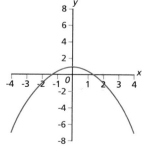

(c)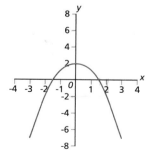

Investigation 3
Changes in Functions, Part 4

Purpose
To develop an understanding of the relationship between the basic graph of a quadratic function and the algebraic change in the equation of the graph of the function

Procedure

A. Construct a table of values for each of the following functions. Use y-values between 0 and 16 such that y is an integer for each table. (Hint: Most y-values will give 2 values for x.)

   (i) $y = x^2$ (ii) $y = (x + 3)^2$ (iii) $y = (x - 4)^2$

B. Refer to the tables of values that you constructed in step A. For the same y-values, how do the x-values for $y = (x + 3)^2$ compare with $y = x^2$? How do the x-values for $y = (x - 4)^2$ compare with $y = x^2$?

C. Graph the three functions of step A. What do you notice about the graphs? How do the graphs compare?

D. Fill in the blanks:

   "When I see $x + 3$ in the equation, then I know that there is a horizontal translation by ▇ units to the ▇▇▇▇."

   "When I see $x - 4$ in the equation, then I know that there is a horizontal translation by ▇ units to the ▇▇▇▇."

   "The horizontal translation is the ▇▇▇▇ of the number being added to the x."

Investigation Questions

16. Describe the transformational relationship between each pair of equations in words and in mapping notation.

    (a) $y = (x - 1)^2$ and $y = x^2$
    (b) $y = 2(x + 3)^2$ and $y = x^2$
    (c) $-y = (x - 5)^2$ and $y = x^2$
    (d) $-2y = (x - 2)^2$ and $y = x^2$

17. Which equation describes each of the graphs at the top of the next page? How do you know?

    A: $-y = (x + 2)^2$ B: $-2y = (x - 1)^2$ C: $y = (x - 3)^2$

(a) C

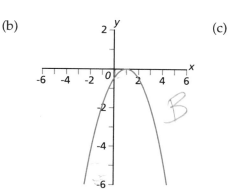
(b) B

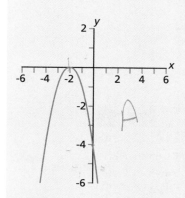
(c) A

Check Your Understanding

18. Each graph below is the result of sliding the function $y = x^2$. Write the equation for each graph.

(a) 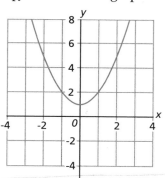 (b)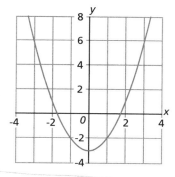

19. Write the equation for the graph of each quadratic function below.

(a) 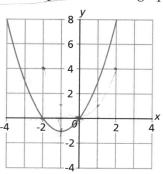 (b)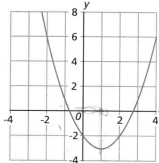

$(x, y) \to (x-1, y-1)$

$y + 1 = (x+1)^2$

$(x, y) \to (x+1, y-3)$

$y + 3 = (x-1)^2$

CHALLENGE yourself

A grasshopper is 80 cm from a wall. With each jump towards the wall, it cuts its distance from the wall in half. The results after the first three jumps are shown in the table.

Number of Jumps	Distance From Wall (cm)
0	80
1	40
2	20
3	10

(a) Are the data discrete or continuous? Plot a graph that shows the total distance from the wall versus the number of jumps.

(b) Which model, linear, quadratic, or exponential, best fits the data? Justify your response.

(c) Write an equation that expresses the distance as a function of the number of jumps.

(d) Use your equation to find the distance the grasshopper would travel in 100 jumps.

20. Describe the translation of $y = x^2$ that has resulted in each image equation in words and using mapping notation.
    (a) $y - 3 = x^2$
    (b) $y + 2 = x^2$
    (c) $y = (x + 5)^2$
    (d) $y = (x - 14)^2$
    (e) $y - 2 = (x + 3)^2$
    (f) $y + 5 = (x - 1)^2$

21. Apply the mapping rule given to the graph of $y = x^2$ and graph the resultant image.
    (a) $(x,y) \rightarrow (x + 1, y)$
    (b) $(x,y) \rightarrow (x, y - 4)$
    (c) $(x,y) \rightarrow (x - \frac{1}{2}, y + 3)$
    (d) $(x,y) \rightarrow (x + 6, y - 5)$
    (e) $(x,y) \rightarrow (x + 2, y - 3)$
    (f) $(x,y) \rightarrow (x - \frac{3}{2}, y - \frac{7}{2})$

22. Sketch the image graph of $y = x^2$:
    (a) with a reflection in the x-axis followed by a vertical translation of 2 units down.
    (b) with a vertical stretch of 4 and a horizontal translation of 3 units to the left.
    (c) given by the equation: $2(y - 1) = (x - 1)^2$.
    (d) given by the equation: $-\frac{1}{2}y = (x + 2)^2$.

23. Describe the transformation of the graph of $y = x^2$ for each graph below. Write the transformation first in words, then as a mapping rule, and finally as an equation.

    (a)
    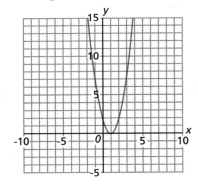

    (b)
    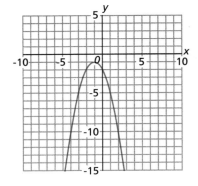

Think about...

Question 21

How can these questions help you sketch a graph?

- Does the graph open up or down?
- Where do you see the vertex?
- Do you see any patterns? symmetry?

— Note —
Reflection in the x-axis is visible in the equation with a $-y$.

— Note —
When there is no stretch factor, the pattern for plotting a quadratic graph is
over 1 up 1
over 2 up 4
over 3 up 9
Starting at the vertex. How would you change the pattern for $\frac{1}{3}y = x^2$?

Absolute Value Functions

Heights and depths can be written using integers. For example, in the earth's atmosphere, the troposphere extends to a boundary 18 km above the earth's surface and its height can be written as +18 km. The earth's crust extends to about 18 km below the earth's surface and its distance from the earth's surface can be written as –18 km. The boundaries are both 18 km away from the earth's surface but are in opposite directions.

Numbers on a number line work in the same way. The absolute value of a number represents its distance from zero on a number line, not showing whether the number is negative or positive.

The symbol $|x|$ means *the absolute value of x*. The absolute value of x is never negative.

Focus Questions

24. Construct a table of values for $y = |x|$. Use x-values between –4 and 4 such that x is an integer. Graph the function. In your own words, describe the shape of the function.

25. Predict the shape of the graph of $-y = |x|$. Check your prediction on a graphing calculator.

26. Describe the relationship between the graphs of the two functions in Questions 24 and 25.

27. Describe in words how the graphs in parts (ii) and (iii) compare with the graphs in part (i)

   A: (i) $y = |x|$ (ii) $y = |x| + 1$ (iii) $-y = |x| + 1$

   B: (i) $y = |x|$ (ii) $2y = |x|$ (iii) $\frac{1}{3}y = |x|$

   C: (i) $y = |x|$ (ii) $y + 2 = |x|$ (iii) $y - 3 = |x|$

   D: (i) $y = |x|$ (ii) $y = |x + 3|$ (iii) $y = |x - 4|$

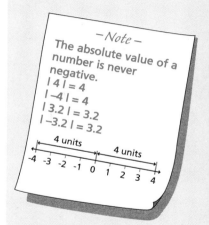

– Note –
The absolute value of a number is never negative.
$|4| = 4$
$|-4| = 4$
$|3.2| = 3.2$
$|-3.2| = 3.2$

– Note –
Repeating the steps and parts of Investigation may help you to complete Questions 27 and 28.

Think about...

Question 28

How are the patterns observed with absolute value functions like those observed with quadratic functions? How are they different?

28. Graph the functions shown in each part of question 27. What do you notice about the graphs? How do the graphs compare with your descriptions? How do the equations compare?

FOCUS G — Putting It All Together

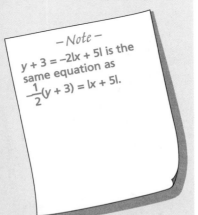

— Note —
$y + 3 = -2|x + 5|$ is the same equation as $-\frac{1}{2}(y + 3) = |x + 5|$.

— Note —
Notice that the graph of $y = |x|$ has been translated 3 units UP to give the image graph with equation $y - 3 = |x|$. The graph of $y = |x|$ has also been translated 5 units DOWN to give the image graph with equation $y + 5 = |x|$.

The results of the four parts of Investigation 3 can be summarized as follows:

Vertical Translation

The graph of $y - q = |x|$ is the image graph of $y = |x|$ after a vertical translation of q units.

Mapping rule: $(x, y) \rightarrow (x, y + q)$

The diagram shows the graphs of $y - 3 = |x|$ and $y + 5 = |x|$.

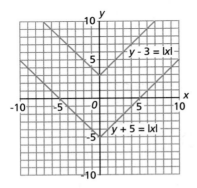

Horizontal Translation

The graph of $y = |x - p|$ is the image graph of $y = |x|$ after a horizontal translation of p units.

Mapping rule: $(x, y) \rightarrow (x + p, y)$

The diagram shows the graphs of $y = |x - 3|$ and $y = |x + 4|$.

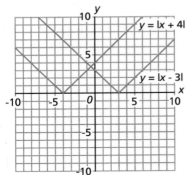

184 Chapter 4 *Modeling Functional Relationships*

Vertical Stretch

The graph of $cy = |x|$ is the image graph of $y = |x|$ with a stretch factor of $\frac{1}{c}$.

Mapping rule: $(x,y) \rightarrow (x, \frac{1}{c}y)$

The diagram shows the graphs of $\frac{1}{2}y = |x|$ and $2y = |x|$.

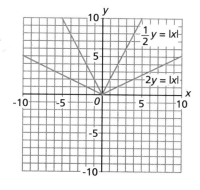

Think about...

Slope after Vertical Stretch

How do the slopes of the two parts of the graph of $cy = |x|$ compare with the value of c?

Reflection in x-axis

The graph of $-y = |x|$ is the image graph of $y = |x|$ after a reflection in the x-axis.

Mapping rule: $(x,y) \rightarrow (x, -y)$

Reflections in the axis often appear with stretches. The graph of $y = -10|x|$, or $-\frac{1}{10}y = |x|$, is shown in the diagram. The mapping rule is $(x,y) \rightarrow (x, -10y)$.

Notice that this graph has been stretched and reflected in the x-axis.

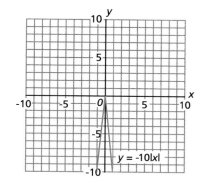

Did You Know?

The processes described here apply to all types of functions, not just absolute value functions. You can, for instance, apply the same reasoning to the graph of $y = x^3$.

The graph of $-\frac{1}{2}(y + 3) = |x + 5|$ is shown. Describe how you know that the graph of this equation will look this way without making a table of values first. Use the graph of $y = |x|$ as a starting point for your description.

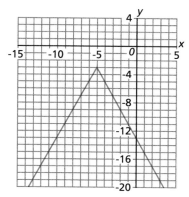

4.3 Equipping Your Function Toolkit

— Note —
When there is no stretch factor, the pattern for plotting an absolute value graph is
over 1 up 1
over 2 up 2
over 3 up 3
Starting at the vertex. How would you change the pattern for $\frac{1}{3}y = |x|$?

CHALLENGE yourself

Sometimes you need to use the skills you used in previous work. Try the following by combining the skills in this section with your skills from Chapter 3.

(a) Make a table of values for lengths and widths of a rectangular garden that has an area of 60 m².

(b) Construct a graph of length versus width from your table of values.

(c) Create a "modeling function" to describe the graph.

(d) Describe the variables used to write your equation. How did knowing the area of the rectangle help?

Focus Questions

29. Refer to the graphs on the previous page, the other graphs that you have drawn in this section, and their corresponding functions. What feature(s) of an equation tell you that each of the following transformations will occur?

    (a) a horizontal translation
    (b) a vertical translation
    (c) a reflection in the x-axis
    (d) a vertical stretch

30. Describe the transformation of the graph of $y = |x|$ that resulted in each graph below. Write the mapping rules and state the equation for each graph.

    (a)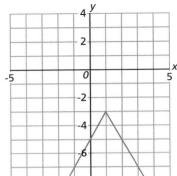
    (b)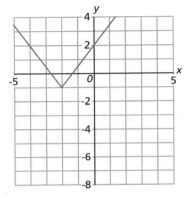

31. Apply the relationship given by each of the following mapping notations to $y = |x|$ and sketch the graph. Write the resultant equation of the graph.

    (a) $(x,y) \to (x + 2, y)$
    (b) $(x,y) \to (x, y - 3)$
    (c) $(x,y) \to (x - 3, y + 6)$
    (d) $(x,y) \to (-x, y + 1)$

32. Draw six different rectangles that are twice as long as they are wide.

    (a) Find the perimeter and area of each of your rectangles. Make a table like this one to record your measurements.

Width	Length	Area	Perimeter

(b) Use your data to make these graphs.
- length as a function of width
- perimeter as a function of width
- area as a function of width

(c) Write functions to represent each of the graphs that you have drawn.

— Note —
Reviewing your work with quadratic equations in Section 3.5 may help you complete Question 32.

Check Your Understanding

33. The graph below represents the heights over time of different balls as they were bounced into the air.

    (a) Write a function for each graph. Explain how you found the function. In your explanation, include:
    - how you used the intercepts;
    - how you used the vertex; and
    - how checking known values helped you to find, modify, and verify the function.

    (b) Write the domain and range for each function.

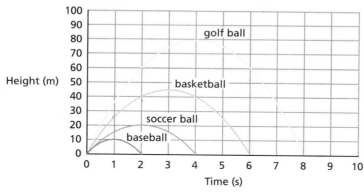

    (c) How could an advertising executive for a toy company use the graph as an advertisement? Is there any information that might help to sell the balls?

CHALLENGE yourself

- Do you think that adding a constant q to any function $f(x)$ will slide the graph of the original function vertically? Why?
- What would happen if you replaced the x in any function $f(x)$ with $x - p$?

Think about...

Each Graph

The graph appears upside down when compared to the graphs of quadratic equations that you explored in Section 3.5. How is this represented in your equation? Why did you decide to do it this way?

34. Light rays were shone onto a MIRA and the path of the light was graphed as shown below. Write a function for each path.

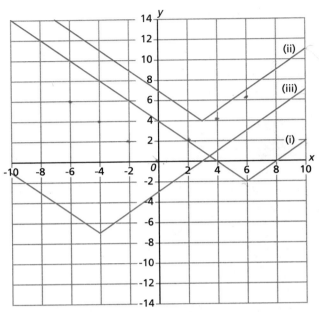

35. A laser beam was reflected off of a mirror and the path of the beam was graphed as shown below. Write a function for each path.

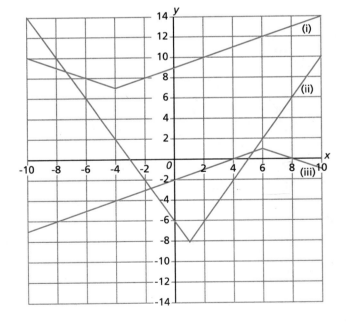

36. Use the patterns that you have investigated in this section to sketch each of the following graphs onto graph paper. Check using technology.

   (a) $2(y - 4) = x^2$
   (b) $y - 3 = |x + 7|$
   (c) $-y = |x - 3|$
   (d) $-(y - 4) = |x + 2|$
   (e) $\frac{1}{2}y = (x - 5)^2$
   (f) $-4(y + 2) = |x - 7|$
   (g) $0.2y = |x + 2|$

37. The graph shown below is $y = x^2$. Kimberleigh, unfortunately, does not recognize the graph and labels it $y = f(x)$ as shown at the right.

   — Note —
   You have probably sketched many of these quadratic graphs throughout Investigation 3. If you have, you can use them here. For each part of this question, you may find it useful to create a table of values and compare the tables of values to help you to sketch the new graphs.

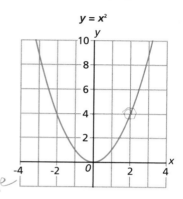
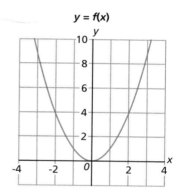

Perform the following transformations on both $y = x^2$ and Kimberleigh's graph, $y = f(x)$.

   (a) Copy the graphs of $y = x^2$ and $y = f(x)$.
   (b) Sketch the graphs of $2y = x^2$, $y = 0.5x^2$ and $y = 0.5f(x)$.
   (c) Sketch the graphs of $y - 3 = x^2$, $y = x^2 + 3$, and $y = f(x) + 3$.
   (d) Sketch the graphs of $y + 2 = x^2$, $y = x^2 - 2$, and $y = f(x) - 2$.
   (e) Sketch the graphs of $y = (x + 3)^2$ and $y = f(x + 3)$.
   (f) Sketch the graphs of $y = (x - 2)^2$ and $y = f(x - 2)$.

What do you notice about the graphs in each part?

4.3 Equipping Your Function Toolkit

4.4 Algebraic Models: Part 1

Manatees are large, gentle sea creatures that live in inlets along the Florida coast. Unfortunately, the manatee is in danger of extinction. Environmentalists would like to see powerboats controlled (in some way) because manatees are often killed or injured by them. In an attempt to analyze the situation, environmentalists gathered data to see if there was a relationship between the number of manatees killed and the number of powerboats. The table in the margin shows the number of registered powerboats (in thousands) and the number of manatees killed in Florida for each year over a fifteen-year period.

Number of Boats Registered (thousands)	Number of Manatees Killed
447	13
460	12
481	24
498	16
512	24
513	20
526	15
559	34
585	33
614	33
645	39
675	43
711	50
719	47
727	54

Investigation 4
Looking for a Trend

Purpose
To see if there is a relationship between the number of manatees that will be killed and a known number of registered powerboats

Procedure

A. Plot the data shown in the margin. Explain why the graph is a more effective way of presenting the data than the table of values.

B. Based on the data, describe any relationships that seem to exist. Describe any trend or pattern in the data.

C. Use your graph to predict the number of deaths likely to occur in one year if the number of registered powerboats reaches 750 000.

D. Check your prediction by estimating the equation for a line of best fit on your data.

Think about...
Step D
A trend line, or line of best fit, can often be used to model the data. How well does the line of best fit match these data?

Investigation Questions

1. When you estimated the line of best fit, did it pass through all of the points? How closely would you say the line fit the data? How did this affect your confidence in the prediction you made using the graph?

2. Using a scale between 0 for *not confident at all* and 10 for *extremely confident*, how confident are you about your prediction? Explain.

Check Your Understanding

3. Three different lines of best fit are drawn for the same scatter plot as shown. One is a line of best fit while the other two represent common mistakes made when the line is drawn. Identify the correct line of best fit and explain the mistakes in the others.

(i)

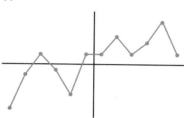

(ii)

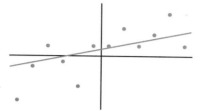

(iii)

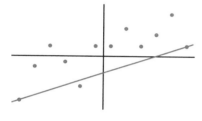

— Note —
You first investigated the line of best fit in Section 1.6. You may need to review that section before trying this investigation.

4. Use a ruler to construct a best-fit line for each graph. Measure the slope and y-intercept and write an algebraic function to represent your best-fit line. How confident would you be in making predictions based on each function? In each case, explain why.

(a)

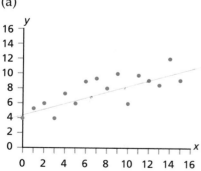

(b)

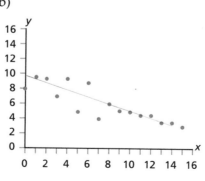

$y = mx + b$

$y = mx + 4$

$y \approx 0.42x + 4$

(12, 9.5)

(0, 4)

$\dfrac{9.5 - 4}{12 - 0}$ $\dfrac{5}{12}$

4.4 Algebraic Models: Part 1   191

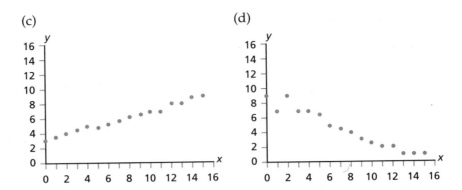

5. Examine the scatter plots in question 4.
   (a) Which plots show a strong correlation? Why?
   (b) Which plots show a less strong correlation? Why?

correlation – a way of describing how well a model (graph or equation) fits the data it represents. If the fit is very close, there is a strong correlation. If the fit is not very close, there is a weak correlation.

FOCUS H The Median-Median Line

If the data seem to show a linear trend, you can produce a best-fit line for the data. Now you will develop a process to find a best-fit line called the median-median line.

Example 1

Two researchers are examining the manatee data in order to make predictions. They want to find the equation of the best-fit line.

Solution

Step 1 Construct a scatter plot for the data.

Step 2 Divide the data points into 3 groups, each containing about the same number of data points. If the number of points does not divide evenly by 3, divide the points so that the outer groups have the same number of points.

Step 3 Find the summary point for each group.

Think about...

The Line of Best Fit

Why is it important to have a method that allows everyone to get a consistent equation to model data?

summary point – one point that best "represents" all the points in a group of data. Its *x*-coordinate is the median of the *x*-values and its *y*-coordinate is the median of the *y*-values.

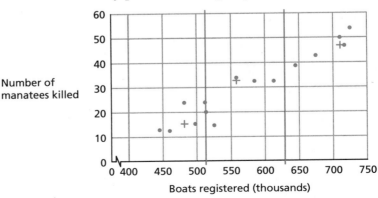

192   Chapter 4 *Modeling Functional Relationships*

Step 4   Draw a faint line through the outer summary points. Find the slope of the line.

Step 5   Move the line in step 4 one third of the way towards the middle summary point, keeping it parallel. This is the median-median line.

Step 6   Find the equation of the median-median line.

Think about...

Step 5
How does this process guarantee that everyone gets the same equation for the best-fit line?

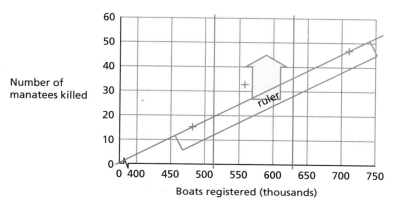

Check Your Understanding

6. These graphs show summary points that have already been calculated as black round dots. Draw the median-median line through these data and find the equation for each median-median line.

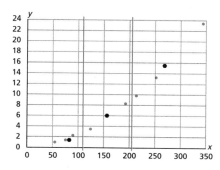
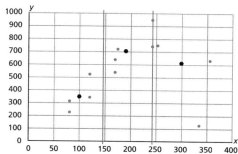

Did You Know?

Graphing technology can be used to find the equation of the median-median line. Enter the points as statistical data and the technology will do the rest!

Group 1	
Time (min)	Amount (g)
3	50
5	200
9	500
11	650
13	800

Group 2	
Time (min)	Amount (g)
15	900
17	890
19	1250
21	1250
23	1600

Group 3	
Time (min)	Amount (g)
25	1600
27	1700
29	1900
31	2000
33	2100

Think about...

Question 9
Compare your predictions in (a) and (b) with those of others in the class. What do you notice? What do you notice about the prediction for the number killed now as compared to the number you predicted in Investigation 4?

Question 10
Is a line the best way to model these data? Give reasons for your answer.

7. The data in the tables in the margin represent time (in minutes) spent picking strawberries and amount picked (in grams). The data have been sorted into three groups of equal size.
   (a) Plot the data points and find the median time and amount for each group.
   (b) Use your answer from part (a). Plot the three summary points.
   (c) Use the summary points to construct a median-median line. Write the equation of the line.
   (d) Suppose your group picked strawberries for 35 min. How many grams do you think you would pick?

8. The following data represent costs for rental cars and distances driven by customers of the Drive-A-Wreck car rental. The data have been sorted into three groups of equal size.

Group 1		Group 2		Group 3	
Distance (km)	Cost ($)	Distance (km)	Cost ($)	Distance (km)	Cost ($)
50	30	100	40	150	55
60	40	110	50	160	50
70	39	120	50	170	60
80	40	130	51	180	60
90	45	140	55	190	65

   (a) Find the equation of the median-median line for these data.
   (b) How much would it cost to drive a car 270 km?

9. Refer to the median-median line constructed for the manatee data in Focus H. Use your equation to predict the number of manatee deaths that will result if 750 000 powerboats are registered.

10. The table below gives the number of grams of fat and the number of calories in 100 g of several salad dressings.

Fat (g)	4.3	42.3	50.2	52.3	60.0	79.9
Calories	96	435	502	504	562	718

   (a) Model the data with an equation. Predict the number of calories in a salad dressing that has 30 g of fat.
   (b) Describe how well your model fits the data. Categorize the correlation as strong, weak, moderate, or non-existent. Give reasons for your choice.

Investigation 5
Picking Margot's Problem

Margot's teacher has given the class five problems. Each student has to choose one problem, solve it, and hand in his or her solution. Margot wants to solve the problem where she has the best chance of getting the most accurate answer.

Problem 1

Ten people each toss a coin for five minutes and count each time a head turns up. These totals are recorded below. If you tossed a coin 45 times, how many heads would you expect?

Number of tosses	26	51	58	32	40	21	50	30	44	47
Number of heads	14	24	24	19	21	14	25	22	20	26

— Note —
These questions will be used in Investigation 5 on the next page. They are presented here for you to read and become familiar with.

Problem 2

Two dice are rolled for five minutes and the number of times a pair turns up is recorded. If you rolled the dice 40 times, how many pairs would you expect?

Number of rolls	42	30	23	56	31	28	54	30	49
Number of pairs	7	4	2	8	10	3	3	5	7

Problem 3

The following data were taken from a study of how monkeys can be trained to perform some tasks. Terry had a monkey for 12 years. How many tasks would you expect the monkey to be able to complete?

Years with person	10.0	8.0	6.5	6.0	5.0	1.5	0.5	0.5
Number of tasks	28	24	28	28	27	23	15	6

— Note —
Data provided in these problems are discrete data. The linear-regression model looks like a continuous line on the screen. Be careful how you interpret and model data.

Problem 4

The number of people entering a theatre to see a movie is recorded for each of the first 10 weeks the movie is showing at the theatre. Suppose you owned the movie theatre and you need 250 people to attend each week to break even or make a profit. For how many more weeks would you run the movie?

Week	1	2	3	4	5	6	7	8	9	10
Number of people	2014	2210	1743	1491	1200	825	1050	1026	998	691

Technology

In this investigation, you will use a line of best fit using linear regression. Therefore, you will need to use graphing technology.

Problem 5

Scientists try to keep track of animal populations to predict how their presence can affect the environment in the area. The number of swans seen at a camp site was recorded for each of 10 straight days. How many swans might there be on the eleventh day?

Day	1	2	3	4	5	6	7	8	9	10
Number of swans	8	18	35	12	54	48	9	26	80	32

Purpose

To decide which set of data will give Margot the most accurate prediction

Procedure

A. Enter data in the tables from the five problems as lists on your calculator. Enter them one problem at a time.

B. Construct a scatter plot for each of the tables you entered in step A. Record the needed information for each problem. Do you think each of these sets of data would be modeled linearly? Why or why not?

C. Which problem is likely to give Margot the most accurate prediction? Give reasons for your answer.

D. Now find and record the linear-regression equation for each of the tables that you have entered.

E. Your calculator gives you a value for r and/or r^2. Record each of these values. Compare these values to your answers in step C. What do you notice?

Investigation Questions

11. Using the values of r and/or r^2, decide which problem Margot should solve.

12. The value of r is called the **correlation coefficient**. Why do you think it is called this?

13. Solve each of Margot's problems in order from the one with the best chance of getting the most accurate prediction, to the one with the least chance of getting the most accurate prediction. Give reasons for the order that you assigned to the problems.

correlation coefficient – a number that describes how well a line models the data. The closer to 1 or –1 the coefficient is, the stronger the correlation. A correlation that is positive means that as one variable increases, the other increases, resulting in a positive slope. A correlation that is negative means that as one variable increases, the other decreases, resulting in a negative slope.

14. Which of Margot's problems gives you a positive correlation? Which gives a negative correlation? Does the correlation being positive or negative influence your order in Question 13? Give reasons for your answer.

Check Your Understanding

15. Write an example of a relation that:
    (a) has a strong positive correlation;
    (b) has a weak correlation;
    (c) has a strong negative correlation; and
    (d) has a zero correlation.

Think about...

Question 15
How might you use Margot's problems as a guide here?

16. Arrange the graphs from strongest to weakest correlation. A line or curve of best fit has already been drawn.

(i) 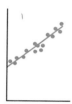   (ii) 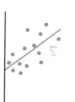   (iii)

(iv)    (v)   (vi)

17. The following table shows the birthrate per 1000 people.

Year	Canada	Nfld	PEI	NS	NB
1975	15.3	23.0	16.5	16.5	17.6
1980	15.5	19.6	16.3	15.7	17.1
1985	14.9	17.9	15.3	14.1	15.0
1990	14.3	14.4	16.0	13.8	13.9
1991	14.1	11.9	14.2	13.0	12.5
1992	13.6	11.4	13.7	12.5	12.3
1993	13.3	10.9	12.8	12.3	11.7
1994	12.9	10.9	12.4	11.9	11.8

— Note —
These data came from the CANSIM data bank of Statistics Canada.

4.4 Algebraic Models: Part 1

Think about...

Question 17
What do the slope and y-intercept of the regression equation represent?

(a) What will the birthrate be in each of the Atlantic provinces in the year 2020? In Canada?

(b) For which province are you most confident in your prediction? Give reasons for your answer.

(c) Which province's birthrate is declining most quickly? Which province should have had the highest birthrate in 1987? How do you know?

18. The data in the table represent amounts of money taken in by a vendor after serving various numbers of customers.

Number of customers	5	10	15	20	25	30	35	40
Amount made ($)	8.75	17.50	27.00	36.75	43.75	52.50	61.25	71.75

Suppose, on Saturday, a vendor anticipates serving 13 customers per hour. How much money can the vendor expect to take in? List any assumptions that you have made.

19. A rowing coach carefully recorded distances traveled in metres after various numbers of strokes by the team. The data are recorded in the table. How many strokes are likely in a 500-m race?

Number of strokes	5	10	15	20	25	30	35	40
Distance traveled (m)	40	60	80	100	110	130	150	170

20. The rowing coach repeated the stroke and distance measurements on another day when there was a cross wind. The data are recorded in the table.

Number of strokes	5	10	15	20	25	30	35	40
Distance traveled (m)	45	65	70	95	100	135	155	160

(a) How many strokes will be needed for a 500-m race?

(b) Find the correlation coefficient for these data. Compare this value with the one from question 19. Account for the difference.

Think about...

Question 21
Do you think that the least squares regression line is the more accurate line? Why?

21. The table gives the mass and oxygen consumption of different harbour seals.

Mass (kg)	25.7	26.8	29.1	35.6	36.8	39.2	41.0	45.2
Consumption (mL/min)	181	266	255	287	301	325	332	344

Estimate the oxygen consumption of a 25-kg seal.

198   Chapter 4 *Modeling Functional Relationships*

4.5 Algebraic Models: Part 2

The table shows the average mass of Atlantic cod as they age. For a person to catch cod for eating, the Ministry of Fisheries determined that the cod must be at least 8 years old. Predict the minimum mass of a cod that can be kept for eating.

Investigation 6
Exploring More Trends Using Technology

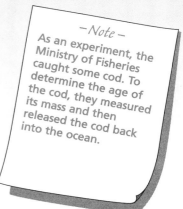

— Note —
As an experiment, the Ministry of Fisheries caught some cod. To determine the age of the cod, they measured its mass and then released the cod back into the ocean.

Age (years)	Mass (kg)
1	0.70
2	1.04
3	1.62
4	2.08
5	3.23
6	4.15
7	6.98
9	11.4

Purpose
To decide on a model that best fits the data, and then to use the model to predict the minimum mass of a cod that can be kept for eating

Procedure

A. Plot the data from the table. Do the data appear to show a definite pattern or trend?

B. Use technology to find an equation for the line of best fit for these data. What is the correlation coefficient? Is this what you would have expected?

C. Use technology to find other equations that can represent the pattern or trend in step A. You should try each of the following:
- a power regression;
- an exponential regression; and
- a quadratic regression.

D. Decide which regression best fits the cod data. How did you make that decision?

Technology
You will need graphing technology to complete this section.

Investigation Questions

1. Use your equation of the regression model from step D. What is the mass for a cod, at least 8 years old, that can be eaten?

2. Suppose you worked in the Atlantic Canada fishing industry. Which model would you want to use to predict the minimum mass of a cod? Give reasons for your answer.

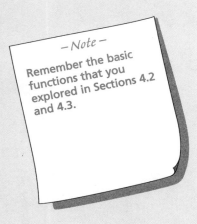

− Note −
Remember the basic functions that you explored in Sections 4.2 and 4.3.

linear: $f(x) = 3x$

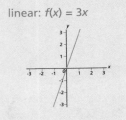

quadratic: $f(x) = x^2$

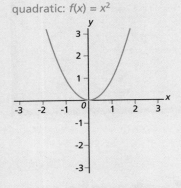

exponential: $f(x) = 2^x$

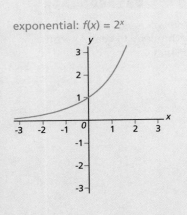

3. Test each of the functions the calculator gave using known pieces of data. How accurate does each function appear to be? How might this process help you decide which function best fits the data given?

Check Your Understanding

4. Use the data in each table to construct a scatter plot. Decide whether the data are best modeled using a linear, power, exponential, or quadratic function. Give reasons for each choice.

(a)

x	5	10	15	20	25	30	35	40
y	600	400	200	100	20	0	30	120

(b)

x	5	10	15	20	25	30	35	40
y	30	25	26	20	24	22	22	20

(c)

x	5	10	15	20	25	30	35	40
y	0.20	0.25	0.40	0.70	1.0	1.7	2.8	4.5

5. Refer to the birthrate data from question 17 of section 4.4 for the four Atlantic provinces.

   (a) Was a linear regression the best way to model the data? If not, what was the best way?

   (b) Use your new model in (a). Which province's birthrate is declining most rapidly?

6. The following data (from CANSIM) show the life expectancy for Canadian males and females.

Year	Canadians	Male	Female	Difference
1920	59.37	58.84	60.60	1.76
1930	61.00	60.00	62.06	2.06
1940	64.58	63.04	66.31	3.27
1950	68.51	66.40	70.90	4.50
1960	71.14	68.44	74.26	5.82
1970	72.74	69.40	76.45	7.05
1980	75.29	71.88	79.06	7.18
1990	77.80	74.61	80.97	6.36

Predict the life expectancy in 2005 for each of Canadians, male, and female, and the difference.

200 Chapter 4 Modeling Functional Relationships

7. Six years ago Avram's parents invested $10 000 in a savings plan for him. The table shows the value of his investment at the end of each year for the past six years.

Year	0	1	2	3	4	5
Balance ($)	10 000	10 700	11 449	12 250	13 108	14 026

Avram is 15 now. How much money will be available to him in three years when he is ready to choose an education or career path?

8. A national construction company is currently selling landfill for $80 per truckload. They estimate that each time they reduce the price by $1 they will get more customers. To test this idea, they sell the fill with various price reductions. The results are shown in the table.

Number of reductions of $1	5	10	15	20	25	30	35	40
Revenue ($)	6750	9100	11 050	12 600	13 750	14 500	14 850	14 800

Find the number of reductions of $1 that results in the most revenue. Explain your method.

9. A balloonist collected the following data as she ascended.

Altitude (m)	0	100	200	300	400	500	600	700	800	900	1000
Temp (°C)	20.0	19.0	18.0	18.0	17.0	16.5	16.0	15.0	14.5	13.7	13.0

Use the equation of the regression model for these data to predict the temperature at 2000 m. How confident are you in your answer?

10. The following data represent the length, in centimetres, of a developing baby. The times are weeks since conception.

Time (week)	8	10	12	14	16	18	20	22
Length (cm)	3.30	7.00	10.50	13.90	17.00	20.00	23.00	25.50

Predict the baby's length at birth (39 weeks). How confident are you in your answer?

PUTTING IT TOGETHER

— Note —
In Chapter 1, you looked at using the mean, median, mode, and standard deviation as summary statistics about a set of data. You may wish to review these ideas before trying this case study.

CASE STUDY 1

The TSE 300 is a measure of how well or how poorly a selection of the top 300 stocks are performing. If the TSE 300 composite index rises, stocks are typically earning money for the investors. If the TSE 300 composite index is doing poorly, stocks are typically losing money for investors. Of course, this is not a guarantee that a particular person's stocks are losing money. It is only an overall indication of how well or how poorly stocks are performing.

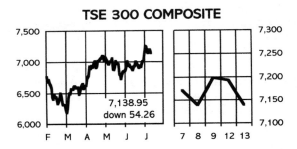

Trading for the last six months and the last five days. Yesterday's close and change from previous.

TSE 300 COMPOSITE

7,138.95 down 54.26

(a) What information is shown by the graphs?

(b) In what career might you be involved if you need this information? How might you use this information?

(c) Why do you think that a broken-line graph is used to display these data? Give reasons for your answer.

(d) What conclusions might you make based on this information? Write a story from the graphs to help support your conclusions.

(e) Suppose you wanted to model these data so that you could make mathematical predictions. Find a function that models this TSE report for the last six months. How confident would you be in making predictions based on your function?

202   Chapter 4   *Modeling Functional Relationships*

EXTENSION 1 BE CAUTIOUS; LIVE LONGER!

You've probably noticed that there is a strong positive correlation between the amount of time that you study for a test and the mark that you receive on the test. Amount of study time is one cause of high test scores. However, correlation does not always imply causation! Read Dr. Friedman's article at the right.

(a) Brainstorm factors that can cause people to live longer.

(b) Do you think that there is only one factor that will cause the correlation Dr. Friedman found to be strong and positive? Could there be others? Give reasons.

(c) Richard found a strong positive correlation between the attendance at hockey games and a team's winning percentage. Might you conclude that winning games causes attendance to increase? Could there be other factors?

(d) Some people consider artificial light a form of pollution because it makes the environment unfit for certain activities such as astronomy. Thinking about cause, effect, and correlation, would you agree with this statement? Give reasons for your answer.

> Being cautious is a key to longevity, according to a 60-year study of more than 1000 men and women. Those who were conscientious as children were 30% less likely to die in any one year than their more "freewheeling" peers.
>
> "We don't really know why conscientious people live longer – it's not as simple as wearing your sweater when it's cold outside." said Dr. Howard S. Friedman, who did the research.

EXTENSION 2

The four Atlantic provinces each selected eleven people to take part in a sea-fishing contest. The number of fish caught by each participant and the total mass of the fish were recorded.

Newfoundland and Labrador

Fish caught	12	4	10	9	11	7	5	3	6	13	8
Total mass (kg)	6.08	4.18	6.83	6.54	9.34	5.45	5.74	2.76	3.32	8.46	7.31

Prince Edward Island

Fish caught	8	9	3	7	4	12	6	11	10	13	5
Total mass (kg)	7.27	7.64	1.60	6.64	3.24	7.24	5.76	7.63	7.76	6.60	4.63

> **– Note –**
> In Chapter 1, you looked at different ways to communicate information about sets of data.

Nova Scotia

Fish caught	7	10	11	4	8	9	12	5	13	3	6
Total mass (kg)		5.27	5.31	6.65	4.23	5.61	7.20	10.00	4.58	7.34	3.89	4.92

New Brunswick

Fish caught	7	7	7	7	7	7	7	18	7	7	7
Total mass (kg)		5.08	4.26	6.21	7.34	6.97	5.54	3.75	10.00	4.06	6.41	5.39

(a) Use the methods below to analyze the data. What do you notice:
  - about the mean number of fish caught in each province?
  - about mean total mass of the fish caught in each province?
  - about standard deviation for the number of fish caught in each province?

(b) Would you have predicted that you would get the measures that you did in (a)? Give reasons for your answer.

(c) Based on the statistics in (a), Huran decided that he could not summarize the data or make many predictions. He decided to try to find the equation of the line of best fit for each province. This will let him make predictions about the mass of fish caught. Find the equation for the line of best fit for each province. What do you notice?

(d) Huran then decided to find the correlation coefficient for each province. He thought that, at the very least, he could determine which line best models the data from the provinces. Find the correlation coefficient. What do you notice?

(e) Based on the data collected, Huran could not make any firm conclusions. He decided to graph the data on a scatter plot. Graph the data. What do you notice? Is it important to graph the data in a scatter plot before you begin to analyze it? Why?

REVIEW

Key Terms

	page
absolute value	183
additive inverse	179
broken-line graph	157
correlation	192
correlation coefficient	196
function	165
horizontal translation	179
mapping notation	175
median-median line	192
multiplicative inverse	178
reflection in *x*-axis	174
regression	195
summary point	192
vertical line test	171
vertical stretch	178

You Will Be Expected To

- interpret tables and graphs.
- recognize linear, quadratic, and absolute value functions and explore exponential and power functions.
- write an equation and calculate using $f(x)$ notation.
- recognize when a relation is a function and when it is not.
- write the domain and range of a function.
- construct a scatter plot and use it to construct a line of best fit using the following methods:
  - "eyeballing";
  - median-median; and
  - least squares regression.
- interpret the correlation coefficient.
- construct a scatter plot and use graphing technology to fit the following curves to data:
  - power; and
  - exponential.
- use transformation techniques in translating among equations, graphs, tables, mapping rules and verbal descriptions.

– Note –
Your story often depends on the assumptions that you make about the situation. All assumptions should be placed at the beginning of the story.

Summary of Key Concepts

4.1 Tables, Graphs, and Connections

Graphs are convenient ways of expressing relationships. Much of the meaning of a graph is conveyed by its slope and *y*-intercept.

- The slope usually shows the rate of change of the dependent variable with respect to the independent variable.
- The *y*-intercept usually refers to an initial condition.

Example

Write a story that describes the graph.

Solution

This graph could represent a skidoo trail.
- The skidoo drives to its first destination at a moderate speed.
- Then it stops there and waits.
- It then heads for its second destination *moving farther away from the starting point, at a greater speed*.
- Finally, it stops at its second destination.

- The horizontal part represents time when the skidoo has stopped.
- Finally, the skidoo returns to the start position.

4.2 Relations and Functions

The shapes of graphs are used to describe a function. A function is a relation for which every value of the independent variable is paired with one and only one value of the dependent variable. It is not a function when one value of the independent variable is paired with more than one value of the dependent variable.

> *— Note —*
> You can use the vertical line test to see, from the graph, whether a relation is a function. If the vertical line intersects the graph at no more than one point at a time, then the relation is a function.

Example

Which picture(s) show functions? How do you know?

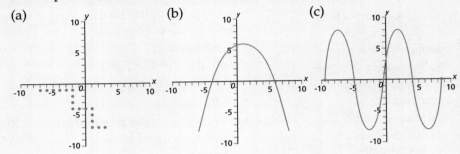

Solution

(a) is not a function because the value of x at -2 is paired with four values of y.

(b) is a function because any value of x is paired with one and only one value of y.

(c) is a function because any value of x is paired with one and only one value of y.

4.3 Equipping Your Function Toolkit

Transformation techniques can often be used to help you to plot a graph. The following is a summary of the patterns that you explored for a function of the form $a(y - k) = (x - h)^2$ and of the form $a(y - k) = |x - h|$.

- Multiplying by a constant a stretches the basic graph $y = x^2$ or $y = |x|$ vertically by a factor of $\frac{1}{a}$.
- Multiplying by -1 reflects the graph in the x-axis.
- Replacing y with $y - k$ results in a vertical translation by k units.
- Replacing x with $x - h$ results in a horizontal translation by h units.

Example

Sketch the function $-\frac{1}{2}(y - 3) = (x - 5)^2$.

Solution

Start with the basic function $y = x^2$.
Reflect the resulting graph in the x-axis.
Stretch the graph vertically by a factor of 2.
Translate the resulting graph by 3 units upward and 5 units to the right.

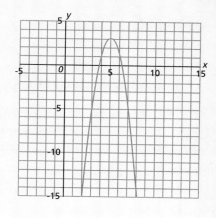

4.4 and 4.5 Algebraic Models

Scatter plots often reveal a linear trend. The correlation coefficient can indicate the strength of the correlation or the goodness of fit. There are several ways by which you can find the equation of the line of best fit.

Example

For a science fair project, a student tested reaction times of people of various ages to auditory and visual stimuli. The data are shown in the tables.

(a) Suppose a 27-year-old person tried the visual stimuli test. About what reaction time would you expect?

(b) Suppose a 21-year-old person took the auditory stimuli test. What reaction time would you expect?

Audio Data

Age	Time (Units)
8	12.786
10	12.333
15	11.378
20	10.678
30	10.036
50	11.825

Visual Data

Age	Time (Units)
8	20.494
10	20.217
15	19.698
20	19.430
30	19.643
50	23.069

Solution

The function for audio stimulus is
$f(x) = 0.005\,061\,9x^2 - 0.31675x + 14.992$, where x represents the age and y represents the audio-reaction time.

The function for visual stimulus is
$f(x) = 0.004\,999\,6x^2 - 0.228\,67x + 22.003$, where x represents the age and y represents the visual-reaction time.

(a) $f(27) = 0.004\,999\,6(27)^2 - 0.228\,67(27) + 22.003$
$= 19.474$

You would expect a reaction time between 19 and 20 time units.

(b) $f(21) = 0.005\,061\,9(21)^2 - 0.316\,75(21) + 14.992$
$= 10.573$

You would expect a reaction time between 10 and 11 time units.

This problem could be looked at using a median-median line or a linear regression line. However, the outlier for 50-year-olds suggests a quadratic equation.

PRACTICE

4.1 Tables, Graphs, and Connections

1. The graph shows temperature readings, in °C, that were taken over a one-day period.

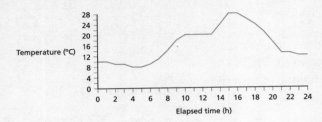

   (a) What is the significance of the slope and y-intercept of the graph?
   (b) What is the significance of the highest and lowest points on the graph?

2. The graph shows the number of meals that were served in each half-hour period during the working day of a restaurant. The business opened at 06:30. Use the graph to "write a story" about the day's activities.

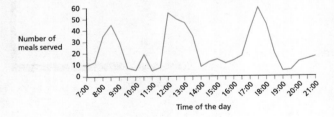

3. Sketch graphs for each description.
   (a) A salmon is swimming upstream. For the first hour, it swims strongly and at a constant speed. For the next two hours, due to fatigue and a stronger current it swims half as fast. It finally realizes it has passed its destination so, for the final half-hour, it stops swimming and allows the current to carry it slowly back downstream.

   (b) By assigning people to watch the exits, a mall manager was able to get a close estimate of the number of people at the mall at various times during the day. Draw a graph to predict the number of people entering the mall on a weekday at one-hour intervals.

4.2 Relations and Functions

4. For each relation, determine if it is a function and explain why it is or why it is not. Classify each graph as linear, quadratic, or exponential.
   (a) cost of catering versus the number of meals served

Number of Meals	Cost of Catering
50	7250
100	9500
150	11 750
200	14 000
250	16 250

   (b) Various locations of a point moving along a path on a plane are represented by the following ordered pairs.
   (0,0), (3,6), (6,24), (6,54), (12,96), (15,50)

   (c) The depth of water in a tank with respect to the time for which water is being discharged is shown in the table below.

Time (s)	Depth (cm)
0	100
20	81
40	64
60	49
80	36
100	25

5. Which relations represent functions, and which ones do not? Give reasons.

(a)

(b)

(c)

(d)

(e)

(f)

6. Suppose that you have groups of people ranging in size from 2 to 6 people. Each person in each group is to shake hands with each other person in the same group.
   (a) How many handshakes are needed in a group of 2? 3? 4? 5?
   (b) Express the information from part (a) in as many ways as you can.
   (c) Explain whether the relation is a function.

4.3 Equipping Your Function Toolkit

7. (a) Graph the function $y + 3 = -(x + 2)^2$. Describe how you did so.
   (b) Graph the function $y + 3 = -|x + 2|$.
   (c) How are your graphs alike? How are they different?
   (d) How was the process you used in (a) like the process you used in (b)? How was it different?

8. Find the equation of the image graph of $y = x^2$ and the equation of the image graph of $y = |x|$ for each mapping relationship shown below. Sketch each graph without constructing a table of values.
   (a) $(x,y) \rightarrow (x - 5, y)$
   (b) $(x,y) \rightarrow (x + 11, -y)$

9. Write an equation to describe each graph.

(a)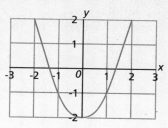

(b)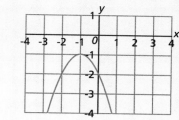

(c)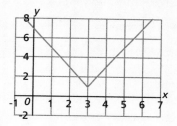

(d)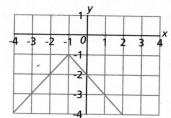

4.4 Algebraic Models: Part 1

10. The lists shows the number of hours of television watched the week just before a test and the test score of 10 students.

    (0,7), (0.5,90), (0.5,95), (1.0,85), (1.5,82)
    (2.0,78), (2.0,8.2), (2.5,75), (3.5,70), (4.0,65)

    (a) Construct a scatter plot from the list.
    (b) Write an equation from the data. Use an appropriate form of $f(x)$ notation.
    (c) Predict the score of a student in the next test who will watch 3 h of television. How confident are you?

11. The graph represents the annual earnings of an employee at various periods of time during his or her employment with the company.

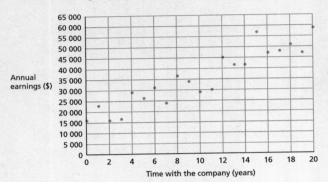

(a) Predict the annual earnings of the employee after the person has been with the company for 25 years.
(b) Explain why you cannot be completely confident in your answer for part (a).

12. Which scatter plot, (a) or (b), gives a regression line that will give the most accurate predictions? Justify your answer.

(a)

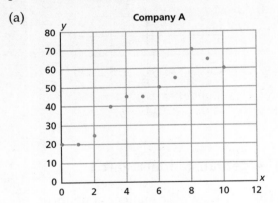

(b)

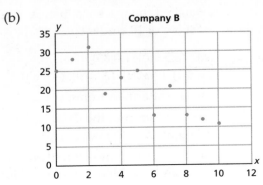

4.5 Algebraic Models: Part 2

13. The table lists the lengths and areas of rectangles that can be produced with 40 m of rope.

Length	2	4	6	8	10	12	14	16	18
Area (m²)	36	64	84	96	100	96	84	64	36

(a) Construct a scatter plot from the table.
(b) Decide whether the data best fit a linear, exponential, or a quadratic model.
(c) Write an equation from the data. Use an appropriate form of $f(x)$ notation.
(d) Predict the length of a side when the area is 40 m².

14. Describe several ways in which you can decide whether a set of data best fits a linear, quadratic, or exponential model.

15. A bakery sells cakes by the kilogram. A Black Forest cake that has a diameter of 20 cm and weighs 1 kg sells for $12. The following table shows some of the diameters of Black Forest cakes of the same height versus their prices.

Diameter of cake (cm)	20	28	35	40	45	53	60
Price ($)	12	24	36	48	60	84	96

Predict the cost of buying a 32-cm diameter Black Forest cake.

16. A study was done on the population of brown trout in a lake. The results for each of the first seven years are given in the table.

Year	0	1	2	3	4	5	6
Population	1500	1800	2200	2600	3100	3700	4500

Predict the population in 10 years.

Chapter Five
How Far? How Tall? How Steep?

People have designed different ways to measure length, height, and steepness indirectly. Indirect measurement is often the most efficient and sometimes the only way to measure. Many of these indirect methods make use of the relationships in right triangles.

In this chapter, you will learn how to use the relationships between the sides and angles of right triangles to measure indirectly.

After successfully completing this chapter, you will be expected to:

1. Measure indirectly using similar right-triangle relationships.

2. Understand proofs of the Pythagorean theorem for right triangles.

3. Apply the Pythagorean theorem to investigate square roots.

4. Understand the definitions of the trigonometric ratios for sine, cosine, and tangent, and use them to measure indirectly.

5. Apply trigonometric ratios to solve problems.

5.1 Ratios Based on Right Triangles

Many problems involving indirect measurements can be solved by using mathematical models based on right triangles.

Focus A: Modeling Situations Involving Right Triangles

A helicopter is searching for a stolen vehicle. The helicopter is flying at a speed of 60 km/h at an altitude of 160 m. A spotlight at the bottom front of the helicopter lights up an area on the ground.

Think about...

The Diagram
What does segment $\overline{BC}$ represent?

A policeman asks the pilot for how long a particular spot would be illuminated as the helicopter passes by. Susan thinks that she could find out by drawing the diagram to the right to model the helicopter situation.

Focus Questions

1. Study Susan's diagram. What happens to $\overline{BC}$ if the light points farther ahead? What is the effect on $\overline{BD}$? Use diagrams to support your answer.

2. Suppose the helicopter flew more closely to the ground. What effect will this have on $\overline{BC}$? On $\overline{BD}$? Use diagrams to support your answer.

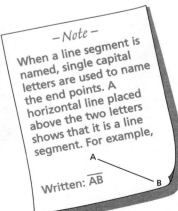

— Note —
When a line segment is named, single capital letters are used to name the end points. A horizontal line placed above the two letters shows that it is a line segment. For example,

Written: $\overline{AB}$

212   Chapter 5   *How Far? How Tall? How Steep?*

Congruence and Similarity

Two triangles are congruent if all the angles and all the side lengths of one triangle match all the angles and all the corresponding side lengths of the other. It looks as if the first triangle had been moved to a new spot.

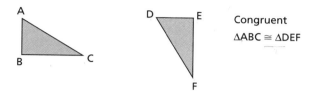

Congruent
△ABC ≅ △DEF

— Note —
When naming a triangle, single capital letters are assigned to the three vertices. The three letters in any order, preceded by a △ symbol, are used to represent the triangle.

Two triangles are similar if one triangle is an enlargement or reduction of the other. The triangles have exactly the same angles, but each of the side lengths of one triangle is the same multiple (could be a decimal number such as 1.5) of the corresponding side length of the other.

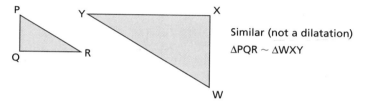

Similar (not a dilatation)
△PQR ~ △WXY

— Note —
A dilatation is a transformation which enlarges or reduces a shape but does not change its proportions. Similarity is the result of a dilatation. The similar quadrilaterals in this diagram below show a dilatation.

Investigation 1
Examining Ratios in Scale Drawings

Purpose
To investigate ratios of side lengths in right triangles.

Procedure

A. Create a scale drawing of △ACD from Susan's drawing in Focus A. In your drawing, let 1 cm represent 20 m.

B. Draw three line segments parallel to the base inside your triangle. Each line segment should be 2 cm apart. Label the vertices as shown. Use this diagram as a guide only.

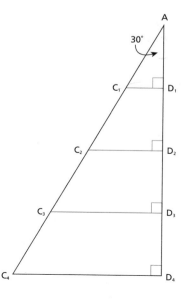

Think about...

Step B

Why do the four triangles △ACD have to have the same angle measurement? Are the triangles congruent? Similar?

5.1 Ratios Based on Right Triangles 213

C. For each triangle in the scale drawing, measure to the nearest tenth of a centimetre the length of the side opposite to $\angle A$, the length of the side adjacent to $\angle A$, and the hypotenuse length. Organize your data in a table like the one below.

Triangle Number	Measures			Ratios		
	Opposite to $\angle A$	Adjacent to $\angle A$	Hypotenuse	$\dfrac{\text{Opposite}}{\text{Hypotenuse}}$	$\dfrac{\text{Adjacent}}{\text{Hypotenuse}}$	$\dfrac{\text{Opposite}}{\text{Adjacent}}$
1	1.2 cm	2.0 cm	2.3 cm			
2						
3						
4						

D. Repeat steps A and B, changing $\angle A$ to 45° (to match Susan's $\triangle ABD$ from Focus A).

Investigation Questions

3. Can you add the same length to all sides of a triangle of your choice and be sure that the new triangle is similar to the original one? Explain.

4. Choose two triangles from step B in which the opposite side to $\angle A$ in the second triangle is double the length of the opposite side to $\angle A$ in the first triangle.
   (a) If you double the height of a right triangle, what happens to the length of the hypotenuse? The length of the base?
   (b) Which triangle measurements did you use to confirm your answer to part (a)?

5. What do you notice about the ratios you formed in step C?

— Note —

The longest side in a right triangle is the **hypotenuse**. Each side $\overline{CD}$ is **opposite** to $\angle A$; notice that it is the only side in each right triangle that does not touch $\angle A$. Each side $\overline{AD}$ is **adjacent** to $\angle A$; it is not the hypotenuse and it touches $\angle A$.

Think about...

Question 5
Why do some people think of similar triangles as triangles in which one triangle is an enlargement of the other?

Check Your Understanding

6.

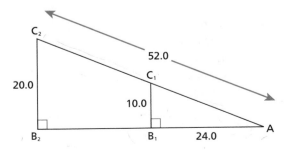

214   Chapter 5 *How Far? How Tall? How Steep?*

(a) Which triangles are similar and how do you know?
(b) How long is $\overline{AB_2}$? How do you know?
(c) How long is $\overline{AC_1}$? How do you know?

7.

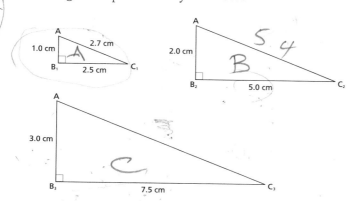

(a) Are the three right triangles similar? How do you know?
(b) Find the perimeter and area of each triangle.
(c) Compare the perimeters between any two triangles. How does the change in perimeter relate to the ratio of the sides?
(d) Compare the areas between any two triangles. How does the change in area relate to the ratio of the sides?
(e) Predict what would happen to the perimeter and area if you multiplied the dimensions of a triangle by 5. Test your prediction.

8. A right triangle has a hypotenuse of 10 m and one leg of 6 m. Another right triangle has two equal legs, each of 30 m. Could the triangles be similar? Explain.

9. To find the height of a tree, you can compare the shadow of the tree to a person's shadow and use the properties of similar triangles.

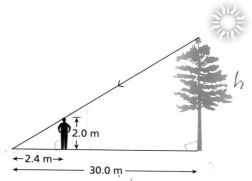

— Note —
The perpendicular sides of a right triangle are sometimes called the legs.

(a) Where are the similar triangles?
(b) A tree casts a shadow that is 30.0 m long. At the same time, a 2.0 m person casts a shadow that is 2.4 m long. How tall is the tree?

10. Another way to find the height of a tree is to use a mirror. You place the mirror on the ground so that when you look down at it, you see the top of the tree.

— Note —
The angle at which light hits the mirror is equal to the angle of reflection.

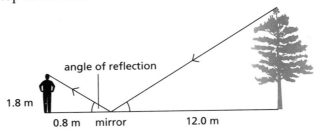

(a) How can you show that the two triangles in the picture above are similar?

(b) Suppose you were 1.8 m tall and you stood 0.8 m from the mirror to see the top of the tree. The base of the tree is 12.0 m from the mirror. How tall is the tree?

11. (a) How could you use the information you gathered from Focus A to find the length of $\overline{CD}$?

(b) How could you use what you know about right triangles with angles of 45° to tell the length of $\overline{BD}$?

(c) Use the results above to calculate:
   (i) the length of $\overline{BC}$
   (ii) the length of time point B would be illuminated.

Vectors and Bearings

In real-life situations, like hiking, camping, or sailing, people use compass directions, along with a distance, to tell where to find a location or object.

There are also other systems used to describe directions and distances.

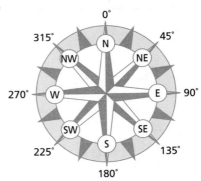

Did You Know?

The Canadian Orienteering Federation was founded in 1967. Nova Scotia was one of the three founding members. Newfoundland and New Brunswick joined in the 1970s. Information about the sport is available on the Federation's web page. (http://www.orienteering.ca)

Sometimes, people use **bearings** with a distance. A bearing tells the direction using an angle measured clockwise from north. An easterly direction is read as 90° and a westerly direction is 270°. Compass bearings are based on magnetic north.

bearing – the angle of direction clockwise from north

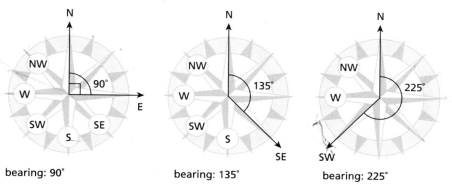

bearing: 90°   bearing: 135°   bearing: 225°

— Note —
This focus assumes that you are working with magnetic north.

At other times, people use **vectors**. Vectors are drawn as arrows. Each shows a distance and a direction. A scale is needed when vectors are drawn. In the case below, 1 cm represents 1 km.

vector – an arrow that shows both direction and distances

For example, all three of the following describe the same distance and direction:

- 5 km east
- 5 km at a bearing of 90°
- 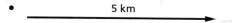 5 km

Did You Know?

Airport runways can be named using bearings. For example, a runway which would be approached from the southwest at a bearing of 32° would be named Runway 3. The name comes from rounding 32 to the nearest ten and dropping the zero. If the runway were approached from the northeast instead, the same runway would be named Runway 21 since the bearing would then be 212°.

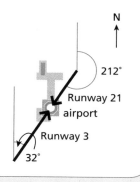

Check Your Understanding

12. Describe these vectors as compass directions and distances, and then as bearings and distances.

(a)   (b)   (c)

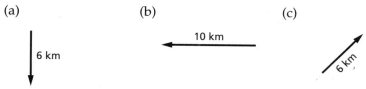

5.1 Ratios Based on Right Triangles   217

13. Draw these directions as vectors. 1 cm should represent 1 km.
  (a) 4 km at a bearing of 90°
  (b) 6 km at a bearing of 45°
  (c) 3 km at a bearing of 135°

14. (a) Which vector could describe 80 km NW?

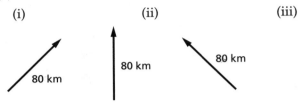

(i)   (ii)   (iii)

(b) Which vector could describe 80 km at a bearing of 30°?

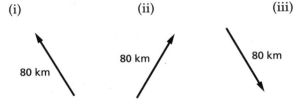

(i)   (ii)   (iii)

15. Mark is walking 5.0 km/h on his orienteering course. The instructions he picks up read:

> Head with a bearing of 90° for 3.0 km.
> Head with a bearing of 225° for 4.0 km.

The first part of Mark's path is shown with a vector.

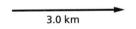

3.0 km

1 cm represents 1 km.

(a) How much time would it take Mark to complete the path required by his instructions?
(b) Figure out where Mark will finish. Use a diagram to help you.
(c) How far apart are the start and finish locations? Write a set of instructions that would get Mark there more quickly.
(d) How much time would it take to go the shorter way?

16. Jane and William started walking from the same spot.
Jane headed at a bearing of 90° for 3.0 km.
William headed at a bearing of 180° for 4.0 km.

(a) Use a scale drawing to find out how far apart Jane and William are after they have followed their instructions.
(b) What instructions should Jane follow to get to William now?

Chapter Project

Planning a Roof

Introduction

Have you ever thought about the different possible designs for house roofs? For this project, you will be creating the design to plan the roof for a house. The roof can be different sizes. The plan for this roof is a triangular prism with isosceles triangle bases.

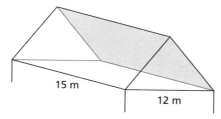

(a) Your first plan must use a scale drawing with bases which are isosceles triangles. Choose any angle you wish for the top angle of the isosceles triangles. Record the measure of the angle.

(b) Draw this angle accurately on paper. Use a convenient scale for the 12-m length. Then use your scale drawing to find the actual height of the roof.

5.2 The Pythagorean Theorem

Suppose the pilot was asked to estimate the length of the light beam in the helicopter search problem. He could use the right triangles in Susan's drawing in Focus A, page 212, along with the Pythagorean theorem, to figure it out.

FOCUS D: Interpreting the Pythagorean Theorem

The Pythagorean theorem states that in any right triangle, the sum of the squares of the two legs equals the square of the hypotenuse. It is often written as:

$$a^2 + b^2 = c^2$$

Visually, the areas of the squares on the sides total the area of the square on the hypotenuse.

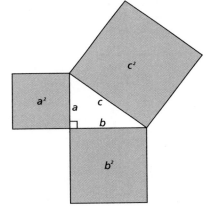

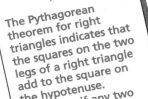

— Note —

The Pythagorean theorem for right triangles indicates that the squares on the two legs of a right triangle add to the square on the hypotenuse. Therefore, if any two sides are known, the third side can be found.

For example,

$3^2 + 4^2 = 5^2$ since $9 + 16 = 25$

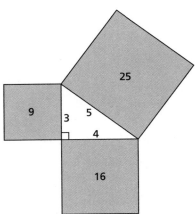

Check Your Understanding

1. Find the missing side lengths.

   (a)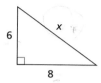

   (b)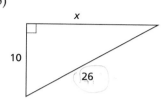

2. You have already seen that 3 cm, 4 cm, and 5 cm are the sides of a right triangle. The Egyptians used this triangle relationship to test if corners were really square in their constructions. They used a rope and tied 12 even knots along the rope. Then they would lay 3 knots along one side of the angle and 4 knots along the other to see if the third side was actually 5 knots long.

   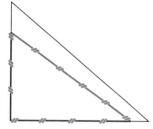 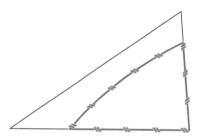

   Make your own knotted rope. You can use non-stretchy heavy yarn and simply mark a dark line at each unit.

   Use it to test several right angles, for example, the corners of the classroom.

3. You might not have realized it, but the Pythagorean theorem is often used in daily life.

   (a) You want to use a sharp cutter edge to cut a small pile of standard looseleaf paper from corner to corner to form triangles for a project. A regular ruler is not long enough to measure accurately across the diagonal of a sheet. How could you find the length of the diagonal to be cut if you can only measure

> **Did You Know?**
>
> Pythagoras was born about 580 B.C. on the island of Samos, but studied in Egypt. He founded the School of Pythagoras in Southern Italy. Math involving geometry, music, and astronomy was studied at the school. He is best known for the Pythagorean theorem relating the side lengths in right triangles. The theorem is named after Pythagoras, even though some people believe that he was not actually the first person to notice this relationship in right triangles.

accurately the length and width of a sheet? What would be the minimum length of a cutter edge you should use?

(b) A case for carrying a musical instrument is 50.0 cm long and 20.0 cm wide. Would an instrument that is 55.0 cm long fit inside the case?

(c) A stripe running from corner to corner in a rectangular room is 5.0 m long. What do you think the length and width of the room might be?

Investigation 2
Proving the Pythagorean Theorem

Purpose
To find out how the Pythagorean theorem can be proved

Procedure
Method 1:

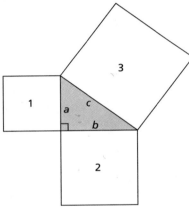

Did You Know?

James A. Garfield, who served as the President of the United States for less than a year before he was assassinated, has been credited for inventing a proof of the Pythagorean theorem based on the area formula for trapezoids.

A. Given a copy of the triangle and squares above, cut out each of the three squares and the triangle.

B. Cut out three additional copies of the right triangle. Label the sides a, b, and c of each triangle.

C. Lay out the four triangles as in the diagram.

Trace the four-sided figure in the middle. How can you prove that it is a square?

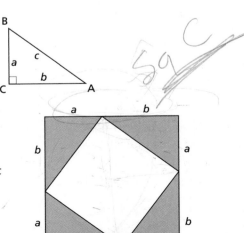

222 Chapter 5 How Far? How Tall? How Steep?

D. Which square from step A fits in the empty space in the middle of the large square in step C? Why does this show that $(a+b)^2 = 4 \times$ area of $\triangle ABC + c^2$?

E. Take the other two squares from step A and combine them with the four triangles that you cut out to form a square with a side length of $(a+b)$. Why does this show that $(a+b)^2 = 4 \times$ area of $\triangle ABC + a^2 + b^2$?

F. How can you combine steps D and E to prove that $a^2 + b^2 = c^2$?

Method 2:

G. Copy this diagram. Given a diagram like the one on the right, cut out the five pieces making up the square on the hypotenuse.

H. Find a way to place the five pieces on the two squares marked A and B that completely cover them with no overlap.

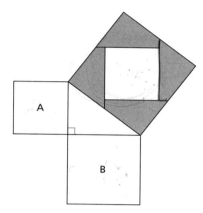

Method 3:

I. Given the right triangle $\triangle ABC$, notice that there are two other triangles in the diagram. Determine which pairs of the three triangles are similar. You may want to redraw the given diagram as three separate triangles.

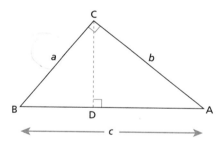

J. Set up proportions based on the lengths of corresponding sides of these similar triangles. Use the proportions to show that $a^2 + b^2 = c^2$.

Think about...

Step H
Discuss whether Method 2 is a proof of the Pythagorean theorem.

Think about...

Step J
Is Method 3 a proof of the Pythagorean theorem? Why?

Investigation Questions

4. Which method did you find easiest to understand? Why is it easier?

5. Which of the methods are proofs? Which are not? Explain.

Check Your Understanding

6. Choose one of the methods in Investigation 2. In your own words, write out an explanation of how it works.

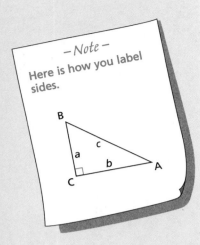

— Note —
Here is how you label sides.

7. Pythagoras proved that in right $\triangle ABC$ where C is the right angle, $a^2 + b^2 = c^2$.
   (i) Is it true then that if $a^2 + b^2 = c^2$, the triangle has to be right?
   (ii) If $a^2 + b^2 \neq c^2$, then does that mean the triangle cannot be right?

   Create some of your own triangles, some of which are right and some of which are not. Measure the three sides of each of your triangles. Label the shortest side "a," the next one "b," and the longest side "c." What conclusions can you draw about (i) and (ii)?

8. Use resources at the library, or the Internet, or other computer search tools for other proofs of the Pythagorean theorem. Bring in one proof that you found. Explain the main idea behind the proof.

FOCUS E — Pythagorean Triples

Did You Know?

If you multiply the three numbers of a Pythagorean triple, the product is always divisible by 60. Why do you think this is so?

If the side lengths of a right triangle are all whole numbers, then these three lengths form a Pythagorean triple. To test if three numbers form a Pythagorean triple, you square the three numbers and determine if the sum of two of the squares equals the third square.

For example,
3-4-5 is a Pythagorean triple since 3, 4, and 5 are whole numbers and $3^2 + 4^2 = 5^2$.

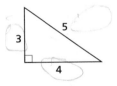

Check Your Understanding

— Note —
3-4-5 is a way to represent a Pythagorean triple, the three side lengths of 3, 4, and 5 of a right triangle.

9. Which of these sets of three numbers are Pythagorean triples?
   (a) 3-4-6
   (b) 4-5-6
   (c) 5-7-9
   (d) 5-12-13
   (e) 7-10-20
   (f) 7-24-25
   (g) 12-35-37

10. Choose one of the Pythagorean triples you found in question 9.
    (a) Multiply each number by 2. Test if the new numbers also form a Pythagorean triple.
    (b) Now multiply each of the original numbers by 10. Test if these new numbers also form a Pythagorean triple.
    (c) If you multiply all three numbers in a Pythagorean triple by the same whole number, do you think you will always get another Pythagorean triple? Why or why not?

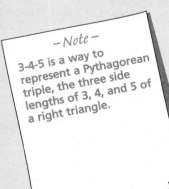

11. Find a pair of possible side lengths for the right triangles shown.

(a)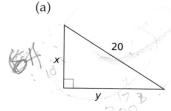

(b)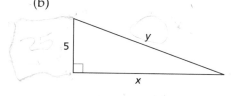

12. Kayoe and Tighe met for lunch in a restaurant. After lunch, Kayoe drove for 9 km at a bearing of 0° back to work. Tighe arrived at work after driving 12 km at a bearing of 270° from the same spot. Kayoe then remembered that she had to pay Tighe for the lunch.

    (a) How far apart are Kayoe and Tighe? Can Kayoe get the money there by driving that far?
    (b) How could you have predicted this knowing that 3-4-5 is a Pythagorean triple?

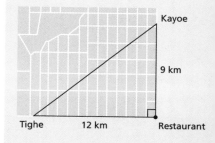

13. The helicopter is flying at an altitude of 160 m with its search light on. The search light has been set to produce a narrow beam, like a laser beam, but much brighter. A rock 160 m is lit up away from the point on the ground directly below the helicopter. What is the length of the beam of light that shines on the centre of that rock?

Chapter Project

Planning a Roof

Continuation

Draw a new pair of triangular bases for the prism forming your roof. This time, check to see that the two right triangles formed inside each isosceles base have side lengths that are Pythagorean triples.

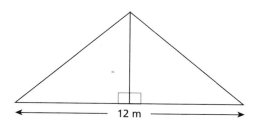

CHALLENGE yourself

The number 24 is one of the numbers in four sets of Pythagorean triples:

18-24-30; 7-24-25; 10-24-26; 24-32-40.

Find another number which is one of the numbers in at least five sets of Pythagorean triples. List the triples to which the number belongs.

5.3 Square Roots and Their Properties

When the Pythagorean theorem is applied to solve real-life problems, very often the side lengths of the right triangle are not whole numbers.

For example, in △ABC
$AC^2 = 6^2 + 4^2 = 36 + 16 = 52$.

Since 52 is not the square of a whole number, the length of the hypotenuse $\overline{AC}$ is not a whole number. You will encounter this type of number in a variety of situations in which you are finding lengths when using trigonometry or solving equations. You need to be able to use, visualize, and interpret these numbers.

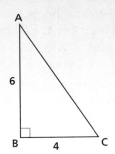

FOCUS F: Square Root Notation

The square root of 16 is 4 or –4. Notice that any whole number has two square roots — one positive and one negative. The positive root is called the **principal square root**.

The symbol $\sqrt{n}$ can be read as "the principal square root of n" or "radical n." It is the positive number which can be multiplied by itself to result in a product of "n."

principal square root – the positive number which, when multiplied by itself, results in the original number

Think about...
Square Roots
Can $\sqrt{n} > n$? Why?

Investigation 3
Relating Square Root to Sum of Squares

Purpose
To represent the square root of a number as a distance on a grid by expressing the number as a sum of two squares

Procedure

A. Draw six different-sized squares on a 5 × 5 square-dot grid. Include no more than three squares in which the sides are horizontal and vertical.

Think about...
Step A
How many non-congruent squares can you make using 5 × 5 square-dot grid?

226 Chapter 5 *How Far? How Tall? How Steep?*

B. Choose one of your squares with slanting sides such as QR. Build a large square around it as shown.

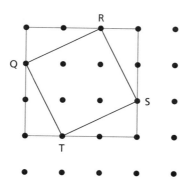

C. Apply the Pythagorean theorem to one of the right triangles you have created around your slanting square to find the side length of your inner square.

D. On a 10 × 10 square-dot grid, create a square with an area of 34.

Investigation Questions

1. What would be the side length of a square with an area of 41?

2. How could writing 41 as the sum of two square numbers help you create the square with an area of 41 on the grid?

Think about...

Step D
Does the square with area of 34 have sides not positioned horizontally and vertically on the grid? How do you know?

Check Your Understanding

3. (a) Find the side length of each square using the Pythagorean theorem.

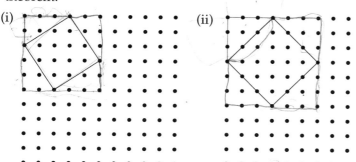

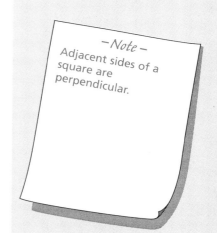

— *Note* —
Adjacent sides of a square are perpendicular.

5.3 Square Roots and Their Properties 227

(iii)

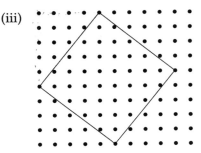

CHALLENGE yourself

State the equation of a line perpendicular to $y = 2x + 7$ having the same y-intercept.

(b) For each square in part (a), start at one of the corners. Count the number of spaces up or down, and right or left, outside the square to get to the next corner. Repeat going from corner to corner.

(i) State the slope of each side of the square.
(ii) State the slopes of any consecutive sides.
(iii) State the slopes of any parallel sides.

Make a conjecture about these relationships.

4.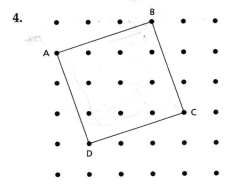

(a) How can you tell from the diagram that the area of the square is less than 16 square units?
(b) How many square units is the area of the square?
(c) Find length $\overline{AB}$ as a square root.
(d) How can you tell from the diagram that AB > 3 units?
(e) How do parts (b) and (d) help you estimate the length of $\overline{AB}$?

5. Write each number as the sum of two squares. Then draw the square root as a length on a square dot grid.
   (a) 52
   (b) 29
   (c) 53
   (d) 45

6. Estimate the square roots below. Tell which can be shown as segments connecting points on a square dot grid. If the segment can be shown this way, check your estimate by drawing the actual length on a grid.
   (a) $\sqrt{32}$
   (b) $\sqrt{65}$
   (c) $\sqrt{59}$

FOCUS G: Simplifying Square Roots

There are many different ways to describe distances using square roots. For example, in the triangle to the right, you might use the Pythagorean theorem to describe AB as

$x = \sqrt{10^2 + 10^2} = \sqrt{100 + 100} = \sqrt{200}$.

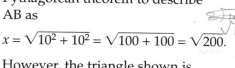

However, the triangle shown is similar to the triangle below it.

Since AC = **10** × 1

and BC = **10** × 1, then

AB = **10** × $\sqrt{2}$

$10\sqrt{2}$ is another way to say $\sqrt{200}$.

Another way to write this is:

$\sqrt{200} = \sqrt{100 \times 2} = \sqrt{100} \times \sqrt{2} = 10\sqrt{2}$

Similarly

$\sqrt{4 \times 2} = 2\sqrt{2}$, $\sqrt{9 \times 2} = 3\sqrt{2}$, $\sqrt{16 \times 2} = 4\sqrt{2}$

This simplification of square roots can also be explained visually by examining the diagrams below.

The area of each little square is 2. The length of each side is $\sqrt{2}$.

You can write: $\sqrt{\text{total area}}$ = length of side in terms of $\sqrt{2}$.

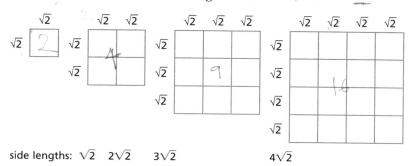

side lengths: $\sqrt{2}$ $2\sqrt{2}$ $3\sqrt{2}$ $4\sqrt{2}$

Whenever the number inside the radical is a multiple of a square number, the square root can be simplified.

For example,

$\sqrt{75} = \sqrt{25 \times 3} = \sqrt{25} \times \sqrt{3} = 5\sqrt{3}$

$\sqrt{98} = \sqrt{49 \times 2} = \sqrt{49} \times \sqrt{2} = 7\sqrt{2}$

— Note —
$\overline{AB}$ is a segment.
AB is its length.

Think about...

The Diagram

The area of the square in the first diagram is 2 with side length $\sqrt{2}$. The area of the second square is 8, so its side length should be $\sqrt{8}$. However, its side length is labeled $2\sqrt{2}$. Are the two lengths equal? Investigate the third and fourth diagrams.

— Note —
When the square root of a whole number is written as a multiple of the square root of a smaller whole number, it is simplified.

Check Your Understanding

7.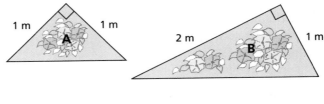

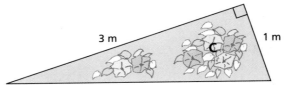

Each length below, measured in metres, is the longest side of a flower bed similar in shape to one of the flower beds A, B, and C in the diagram. To which of A, B, or C is each one similar?

(a) $\sqrt{72}$ (b) $\sqrt{20}$ (c) $\sqrt{32}$ (d) $\sqrt{90}$

(e) $\sqrt{50}$ (f) $\sqrt{200}$ (g) $\sqrt{2.88}$ (h) $\sqrt{\dfrac{20}{25}}$

8. Draw a picture to show why $5\sqrt{8} = \sqrt{200}$.

9. Lydia said she calculated the length of the longest side of her flower bed to be $5\sqrt{3}$ m long. Elijah said that it was actually 8.65 m long by measuring.

    (a) Which length is exact and which is approximate?
    (b) Which length is more useful?

H Adding and Subtracting Square Roots

What does 3■ + 4■ mean? Why is it the same as 7■?

What does $3x + 4x$ equal?

Why should $3\sqrt{5} + 4\sqrt{5} = 7\sqrt{5}$?

In general, what is $a\sqrt{x} + b\sqrt{x}$? Why? $a+b\sqrt{x}$

What is $c\sqrt{x} - b\sqrt{x}$? Why? $c-b\sqrt{x}$

230   Chapter 5 *How Far? How Tall? How Steep?*

Focus Questions

10. Suppose a student wrote this: $2\sqrt{3} + 4\sqrt{2} = 6\sqrt{5}$. Decide whether this is correct. Give reasons.

11. Why might you rewrite $2\sqrt{27}$ as $6\sqrt{3}$ in order to add it to $4\sqrt{3}$? Show with a diagram why $2\sqrt{18}$ and $6\sqrt{2}$ are the same length.

Check Your Understanding

12. Write the sums or differences as a single, simplified square root if possible; if the sum or difference cannot be simplified, explain why.
    (a) $6\sqrt{3} + 8\sqrt{3}$
    (b) $10\sqrt{7} - 6\sqrt{7} - 2\sqrt{7}$
    (c) $3\sqrt{2} + 5\sqrt{8}$
    (d) $5\sqrt{50} - 3\sqrt{2} + 6\sqrt{50}$
    (e) $4\sqrt{5} + 3\sqrt{15}$

    How could you use a calculator to test if your answers are correct?

13. (a) A square of area 50 square units is drawn. A smaller square of area 48 square units is removed from it as shown. Show the dimensions of the remaining piece in simplified form.

    (b) A square room of area 128 m² is used for a concert. A square area of 50 m² at one corner is reserved for seating the band, and the rest of the space is for the audience. Show all dimensions of the room.

    (c) A synchronized swimming performance takes place at an indoor square area of 300 m². The pool area is in the shape of a square located at one corner and occupies a space of 192 m². Show all dimensions of the space left for the audience.

14. (a) Look at your diagram from Focus A, shown again below.

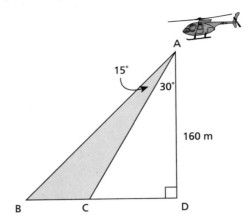

Multiply the length of $\overline{CD}$ you found in question 11 of section 5.1 on page 216 by $\sqrt{3}$. What do you notice?

(b) Form another right triangle similar to $\triangle ACD$. Multiply its base length by $\sqrt{3}$. Does the same thing happen?

CHALLENGE yourself

You add a whole number multiple of $\sqrt{x}$ to $3\sqrt{10}$. The answer is a multiple of $\sqrt{10}$. If x is between 300 and 400, what could x be?

Chapter Project

Planning a Roof

Continuation

To create a comfortable attic, you want to create a design for the triangular ends of your roof so that the height of the roof is at least 2 m.

- First predict the length of the slanting side of the roof end.
- Then describe the length of the slanting side of the end as a simplified square root.
- Estimate the length and then measure to check.

5.4 Defining Trigonometric Ratios

Susan wondered if she really needed to make a scale drawing to find the length of time a spot on land would be lit up by the helicopter. Is it possible to calculate the time if you know the angle spread of the light beam and the height of the helicopter?

Investigation 4
Triangle Ratios

Purpose
To observe relationships in right triangles with the same reference angle

Procedure

A. Use a series of right triangles with a base angle of 40° as shown.

For each of the three triangles, measure the lengths of the three sides to the nearest millimetre. Using ∠A as the reference angle, calculate the three ratios:

$\dfrac{\text{opposite}}{\text{hypotenuse}}$, $\dfrac{\text{adjacent}}{\text{hypotenuse}}$, and $\dfrac{\text{opposite}}{\text{adjacent}}$.

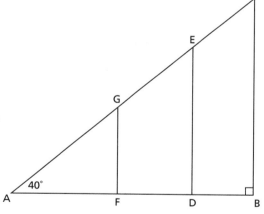

Round all your calculations to the appropriate number of significant digits. Then set up and complete a table in your notebook using the following headings.

Triangle	Opposite to ∠A	Adjacent to ∠A	Hypotenuse	$\dfrac{\text{Opposite}}{\text{Hypotenuse}}$	$\dfrac{\text{Adjacent}}{\text{Hypotenuse}}$	$\dfrac{\text{Opposite}}{\text{Adjacent}}$

B. Create a series of three right triangles like those in step A for each of the following reference angles: 10°, 20°, 30°, 50°, 60°, 70°, 80°. Then set up a chart, listing all the ratios in step A for the three triangles for each reference angle using the headings shown below.

Reference Angle	10°	20°	30°	40°	50°	60°	70°	80°
$\dfrac{\text{Opposite}}{\text{Hypotenuse}}$								
$\dfrac{\text{Adjacent}}{\text{Hypotenuse}}$								
$\dfrac{\text{Opposite}}{\text{Adjacent}}$								

Investigation Questions

1. Are the three triangles that you drew for each reference angle similar? Why or why not?

2. Based on your measurements, what is the appropriate number of significant digits to use to calculate your ratios? Why?

3. For the three triangles with the same reference angle, what did you notice about the ratios of:

   (a) $\dfrac{\text{opposite}}{\text{hypotenuse}}$? (b) $\dfrac{\text{adjacent}}{\text{hypotenuse}}$? (c) $\dfrac{\text{opposite}}{\text{adjacent}}$?

   Are you surprised? Explain why or why not.

4. Use your chart to calculate:

   (a) the length of $\overline{AB}$ and $\overline{AC}$ in the figure to the right; and

   (b) the measure of $\angle A$ if BC remains 50 m and AB is now 18 m.

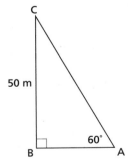

5. A ladder that is 7.9 m long reaches a window 6.1 m above ground. Use your chart to find the angle that the ladder makes with the ground.

Check Your Understanding

6. Find the following information in your tables. For what angle:

   (a) is $\dfrac{\text{opposite}}{\text{adjacent}}$ closest to 1?

   (b) is $\dfrac{\text{opposite}}{\text{hypotenuse}}$ equal to 0.5?

   (c) is $\dfrac{\text{adjacent}}{\text{hypotenuse}}$ equal to 0.5?

   (d) is $\dfrac{\text{opposite}}{\text{hypotenuse}}$ less than $\dfrac{\text{adjacent}}{\text{hypotenuse}}$?

7. As the angle increases, what happens to:

   (a) $\dfrac{\text{opposite}}{\text{hypotenuse}}$?   (b) $\dfrac{\text{adjacent}}{\text{hypotenuse}}$?   (c) $\dfrac{\text{opposite}}{\text{adjacent}}$?

8. Compare each pair of values below. What do you notice? What do the two angle values have in common each time?

   (a) $\dfrac{\text{opposite}}{\text{hypotenuse}}$ for 20° and $\dfrac{\text{adjacent}}{\text{hypotenuse}}$ for 70°

   (b) $\dfrac{\text{opposite}}{\text{hypotenuse}}$ for 30° and $\dfrac{\text{adjacent}}{\text{hypotenuse}}$ for 60°

   (c) $\dfrac{\text{opposite}}{\text{hypotenuse}}$ for 40° and $\dfrac{\text{adjacent}}{\text{hypotenuse}}$ for 50°

FOCUS 1: The Trigonometric Ratios

Since the ratio of opposite side to adjacent side is the same for each right triangle with a given reference angle, this ratio is given a special name. It is called the tangent of the angle and is written as tan X, where X is the reference angle.

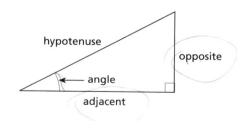

5.4 Defining Trigonometric Ratios

tan X – a constant value based on the ratio of the length of the side opposite to a chosen angle X to the length of the side adjacent to angle X in a right triangle

sin X – a constant value based on the ratio of the length of the side opposite to a chosen angle X to the length of the hypotenuse in a right triangle

cos X – a constant value based on the ratio of the length of the side adjacent to a chosen angle X to the length of the hypotenuse in a right triangle

trigonometric ratios – constant values based on the ratios of sides for particular angles in right-angled triangles

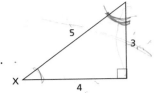

$$\tan X = \frac{\text{length of side opposite to angle X}}{\text{length of side adjacent to angle X}}$$

is read as "tangent of angle X," and equals $\frac{3}{4}$.

Similarly, the other two ratios $\frac{\text{opposite}}{\text{hypotenuse}}$, which represents the "sine of the angle," and $\frac{\text{adjacent}}{\text{hypotenuse}}$, which represents the "cosine of the angle," are also constant for the reference angle X.

$$\sin X = \frac{\text{length of side opposite to angle X}}{\text{length of hypotenuse}}$$

is read as "sine of angle X," and equals $\frac{3}{5}$.

$$\cos X = \frac{\text{length of side adjacent to angle X}}{\text{length of hypotenuse}}$$

is read as "cosine of angle X," and equals $\frac{4}{5}$.

Sine, cosine, and tangent are called **trigonometric ratios**. The word "trigonometry" comes from the Greek words "trigonon" for triangle and "metria" for measure.

Focus Questions

9. How do you know, from the definition of sine, that in a right triangle, the sine of an acute angle is always less than 1? How do you know that the cosine of an acute angle is also always less than 1?

10. How would you define for a classmate what is meant by opposite and adjacent sides?

11. Refer back to the table in step A of Investigation 4. What section in the table tells the sine? the cosine? the tangent?

Check Your Understanding

12. Identify the sides you would use to calculate sin A for each.

(a) 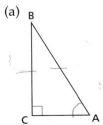 (b) 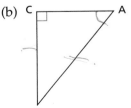 (c)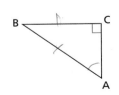

Did You Know?

A line that touches a circle at only one point is called a *tangent*. There is a relationship between this meaning of tangent and the definition of tangent as $\frac{\text{opposite}}{\text{adjacent}}$.

If a circle of radius 1 unit is drawn as shown below and a scale placed beside the tangent line, the value of the tangent of angle *x* can be read directly off the scale at the point where the upper arm of the angle touches the tangent line.

13. (a) Find cos A in each triangle.

   (i) 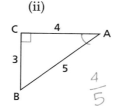 (ii) 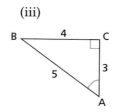 (iii)

   (b) Make up a question similar to part (a) using your own right triangles and challenge a friend to use sine to figure out which angle A is the greatest.

14. (a) Which of the non-right angles below has the greatest tangent?

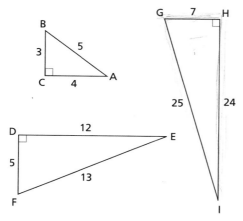

   (b) Which has the least tangent?
   (c) Could you have predicted your answers to (a) and (b) by looking, but not calculating? Explain.

15. Examine the triangle shown.

   (a) Record sin B, cos B, and tan B as fractions, rather than decimals.
   (b) Find sin B ÷ cos B as a fraction. What do you notice?

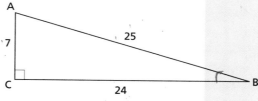

   (c) Create a different right triangle and compute sin B ÷ cos B and tan B. What do you notice? Explain why.

16. (a) Draw a right triangle with a 30° angle. Find the sine of 30°.
   (b) How does the length of the shortest side in this triangle compare to the length of the hypotenuse?
   (c) Examine this diagram of an equilateral triangle. Use the diagram to explain your answer to (a).

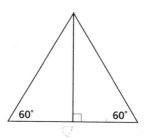

5.4 Defining Trigonometric Ratios

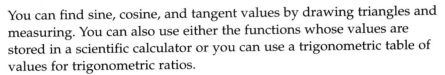

Using a Calculator or Tables to Find Trigonometric Ratios

You can find sine, cosine, and tangent values by drawing triangles and measuring. You can also use either the functions whose values are stored in a scientific calculator or you can use a trigonometric table of values for trigonometric ratios.

To use a calculator, make sure your calculator is in DEGREE mode, and not in radian mode.

Before calculators became readily available, books often included trigonometric tables for people to look up values for sine, cosine, and tangent functions.

Part of a trigonometric table is shown below. The full table is at the back of this text.

Angle in Degrees	Sine	Cosine	Tangent
40	0.6428	0.7660	0.8391
41	0.6561	0.7547	0.8693
42	0.6691	0.7431	0.9004
43	0.6820	0.7314	0.9325
44	0.6947	0.7193	0.9657
45	0.7071	0.7071	1.0000
46	0.7193	0.6947	1.0355
47	0.7314	0.6820	1.0724
48	0.7431	0.6691	1.1106
49	0.7547	0.6561	1.1504
50	0.7660	0.6428	1.1918
51	0.7771	0.6293	1.2349
52	0.7880	0.6157	1.2799
53	0.7986	0.6018	1.3270
54	0.8090	0.5878	1.3764
55	0.8192	0.5736	1.4281
56	0.8290	0.5592	1.4826
57	0.8387	0.5446	1.5399
58	0.8480	0.5299	1.6003
59	0.8572	0.5150	1.6643
60	0.8660	0.5000	1.7321

Focus Question

17. Look at the chart values you calculated in Investigation 4. Compare them to the values for sine, cosine, and tangent you get using your calculator or the trigonometric table. Which uses more digits? Why?

Check Your Understanding

18. Suppose a helicopter light beam is aimed at a 45° angle.

    (a) Find the sine and cosine of angle X using your calculator. What do you notice?

    (b) Why would the tangent be easy to predict?

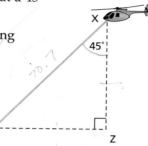

19. Use your calculator to record the sine and cosine values for the following angles: 15°, 30°, 45°, 60°, and 75°.

    (a) What do you notice about the values for 15° and 75°? For 30° and 60°?

    (b) Predict the answers to these and then check.

    (i) $\sin 35° = \cos$ _____

    (ii) $\sin 81° = \cos$ _____

    (iii) $\cos 18° = \sin$ _____

20. (a) Suppose you know that the sine of an angle is 0.8000. What is the angle? Suppose the sine of an angle is 0.8252. What is the angle? What might cause the differences recorded?

    (b) What is the angle if the cosine of the angle is 0.8000?

    (c) Explain how you can find an angle if you know the sine or cosine of that angle.

21. Find the missing side lengths and indicated angle measures in these triangles:

    (a)

    (b)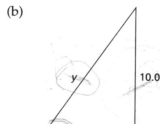

    (c)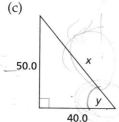

    (d)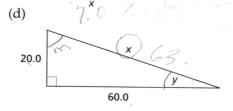

Think about...

Question 20

Would it be easier to use a calculator or a trigonometric table to find an angle if you know the sine or cosine of that angle? Explain.

5.4 Defining Trigonometric Ratios 239

Did You Know?

One of the most famous pyramids in Mexico is the Kukulcán located in Chichen Itza. The steps going up the pyramid are very steep. Recently, visitors have been prohibited from climbing the stairs in order to preserve the stairs.

22. (a) An equilateral triangle has a side length of 10.0 cm. What is its height?
    (b) What will be the area of an equilateral triangle if you know its side length?

23. (a) For safety reasons, a "normal" set of stairs can only have a rise of 72 cm for every 1 m of run. What is the tangent of the base angle B? At about what angle do the stairs rise? How many steps are there per metre?
    (b) The steps on one of the Mayan pyramids in Mexico rise about 64 cm for every 71 cm of run. Is this steeper or less steep than normal stairs? How do you know?

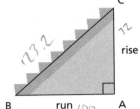

24. Keely noticed a flagpole on the other side of a river, and figured that she could use it to indirectly measure the width of the river. Starting from the point directly across from the pole on the other side of the river, Keely walked 50 m along the bank. According to Keely's compass, the flagpole appeared to be at a 50° angle to her present position. How wide is the river?

25. If the light beam from the helicopter is aimed at an angle of 70° from the vertical as shown below, and the helicopter remains at an altitude of 160 m, what is the distance from the point directly below the light on the helicopter to the farthest point being lit up? If the spread of the light beam is 5°, find the width of the beam on the ground.

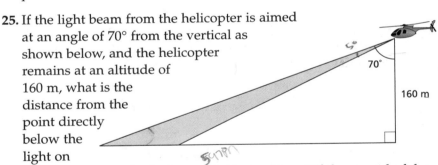

Chapter Project

Planning a Roof

Continuation

For each of the three roof plans you developed in Sections 5.1, 5.2, and 5.3, find the base angle measures for your triangles and compute the tangent of these angles. What do you notice about the base angle of the triangle as the height increases?

5.5 Applications of Trigonometry

Trigonometry helped you find distances and beam lengths in the helicopter problem. It can also be used to solve variations of this problem as well as a large variety of other types of problems involving height, length, and steepness.

Comparing Reaches

Paul Bunyan is the hero of several North American legends. The story goes that he grew to be 149 m tall, about 129 m from shoulder to foot. He was a lumberjack and used his arms to wield axes to cut down trees. The legends say that he dug out the Saint Lawrence River and the Great Lakes. How could you use trigonometry to figure out how long Paul's arms were?

> **Did You Know?**
>
> It is believed that French-Canadian lumberjacks circulated the Paul Bunyan myth. The first written reference to Paul was in a Detroit newspaper in 1910. Paul has been the subject of ballets, plays, and operas and also the hero of many children's books.

Check Your Understanding

1. (a) If the angle from Paul's ankle to his fingertips was 28° from the vertical, how long would his arms be?
   (b) What if the angle was 18°? Which seems more reasonable to you? Why?

2. Hold your thumb out to make an angle of 90° with the side of your index finger. Measure angle X on your own hand and measure the length of your thumb.

   Use trigonometric ratios to predict how far apart Paul's thumbtip and index fingertip might have been, assuming Paul's measurements are about 88 times yours.

3. The shadow of a tree is 20.0 m long. The angle from the ground to the top of the tree is 50°. How tall is the tree?

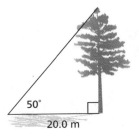

4. (a) A wheelchair ramp for outdoor use generally has a steepness of 1 cm of rise for every 20.0 cm of run.

   If a ramp must reach a door which is 25 cm above the ground, how far from the building must the ramp start? What is the angle of slope of the ramp?

   (b) Indoor ramps generally have a steepness of 1 cm of rise for every 12 cm of run. How far along the floor would an indoor ramp have to be to rise 25 cm? Is this more or less steep than an outdoor ramp? What is the angle of the slope?

5. An auto repair shop builds a ramp so that mechanics can stand below a car to work on it. Assuming a mechanic might be 2.0 m tall, if the angle at which the ramp meets the ground is 20°, how long should the ramp be?

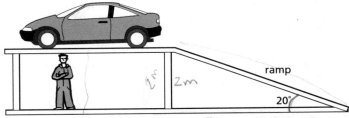

6. (a) How long a ladder would you need to reach a 15 m height if the base of the ladder cannot be more than 5 m from the wall?

   (b) At what angle to the ground would the ladder be at this maximum distance from the wall?

7. You are hiking up a fairly steep hill. The angle of rise of the hill is about 10°. The hill trail is close to straight. After 2.2 h, you are at the top of the hill. Your speed is 4.0 km/h. About how high up from the ground are you at the summit?

8. Jesse is standing in front of three electrical poles which are lined up side-by-side. He is 80.0 m away from the first pole on a line perpendicular to the line of poles. He can see the second pole at an angle of 40° and the third pole at an angle of 55°. How far is the second pole from the third?

9. Maeve and Leighann live in houses on the opposite sides of a building having at its top a two-sided digital display of time and temperature at a height of 22 m. Maeve has to look up from the ground at a 26° angle to see the display and Leighann has to look up at a 35° angle. How far apart do the girls live?

10. Richmond and Keith are 200.0 m apart from one another along the same side of a straight street. They both see their friend's car on the opposite side of the street 100.0 m from Richmond. The car is between Keith and Richmond. The angle at which Richmond sees the car is 50° from the side of the street. At what angle does Keith see the car?

11. (a) A patio in a town park is built in the shape of a regular hexagon measuring 20.0 m straight across. Calculate the perimeter and area of the hexagon.

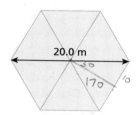

(b) Another patio in a mall is in the shape of a regular pentagon with a side length of 10.0 m. Find the area of this pentagon.

(c) Can you conclude that if the patio is a regular polygon with a side length of 10.0 m, the patio with more sides will have a greater area? Explain.

12. Kevin always wanted an unusual desk. The desk that he has now is a rectangular one which is 130.0 cm × 80.0 cm. In the newspaper, he saw an ad for a desk shaped like a regular pentagon. Each side was 80.0 cm. Is the area of this desk greater or smaller than the one he has? How do you know?

13. A pool table is 1.3 m wide and 2.6 m long. A white ball is shot to rebound and hit the black ball. The angle at which the ball hits the side is the same as the rebound angle. The positions of the balls are shown on the diagram.

Use trigonometric ratios to find the distance the white ball traveled by the time it hit the black ball.

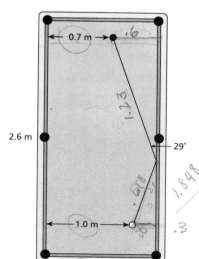

$\tan 30 = \dfrac{10}{h}$

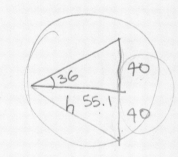

14. Recall the helicopter problem in Focus A at the beginning of this chapter. The problem was set up so that the helicopter was flying at 160 m and the light beams spread between angles of 30° and 45°. How would the length of time that a spot is lit up change if:
    (a) the angles remained the same, but the helicopter flew at an altitude at 120 m?
    (b) the helicopter still flew at an altitude of 160 m, but the beam spread between angles of 40° and 80°?

K Angle of Elevation

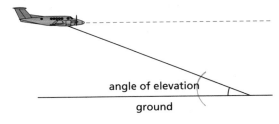

angle of elevation – the angle between the ground and the direction you must look up to see an object

If someone were on the ground looking up at the plane, the angle at which the person must look up to see the plane is called the **angle of elevation**.

Check Your Understanding

15. John holds the string of a kite 1.5 m above the ground and flies his kite with a 25 m string. The angle of elevation is 60°. How high above the ground is the kite?

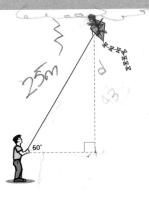

16. Mary, whose eyes are 1.6 m above ground, sees the top of a building at an angle of elevation of 30°. She is 100.0 m away from the building.
    (a) How tall is the building?
    (b) If she were looking up at ground level, what would be the angle of elevation?

244 Chapter 5 *How Far? How Tall? How Steep?*

17. A plane is coming down for a landing. The angle from the horizontal is 25°. The plane is 367 m from its landing point along the ground. How high is the plane?

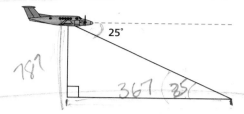

18. From the top of a cliff, the angle from the horizontal looking down toward a boat is 30°. If the cliff is 60 m high, how far away is the boat from the base of the cliff?

19. In many places in the Atlantic provinces, a 15-story building would be very tall.

    (a) Estimate about how tall it would be.

    (b) Suppose you found out that when you looked at the top of the building from one spot, the angle of elevation was 39°, but when you moved 10 m closer, the angle of elevation became 45°. How tall is the building?

20. Chris and Kerry are both looking at the top of the same tree that is 20.0 m tall. Chris is 100.0 m and Kerry is 200.0 m from the base of the tree. They are both on the same side of the tree. Which angle is greater — x or y? By how much is it greater?

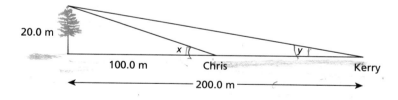

5.5 Applications of Trigonometry 245

Investigation 5
Finding Vector Distances and Directions

In Focus C, you explored orienteering instructions based on bearings and distances. You used vectors to show paths. More complicated paths based on following a number of sets of instructions can be investigated using trigonometric ratios.

Purpose

To use trigonometric ratios to find the result of combining vectors

Suppose that you are following a set of orienteering instructions. You must complete each set of instructions, one after the other, leaving from your most recent position each time. The instructions and resulting diagram are shown below.

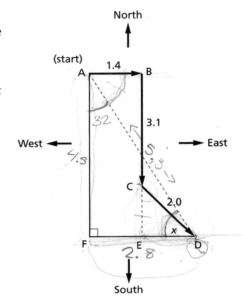

Bearing	Distance	
90°	1.4 km	THEN
180°	3.1 km	THEN
135°	2.0 km	

Procedure

A. Measure angle x. How else could you have found the measure of angle x?

B. Compute $\cos x$. Use that value to find the length $\overline{ED}$.

C. Compute $\sin x$. Use that value to find the length $\overline{CE}$.

D. Describe the lengths of all three sides of $\triangle AFD$. Side $\overrightarrow{AD}$ is called the **resultant vector**. It connects the first and last positions. You can describe it in terms of its length and the bearing you would use to get from A to D directly.

Investigation Questions

21. Is it easy to tell the bearing from A to D based on the orienteering instructions? Give reasons for your answer.

22. How could you find the bearing once you know the total distance south and distance east you traveled? What is the bearing back to A from D?

— Note —
A vector is shown with an arrow over the top of the letters showing the end points.

resultant vector – a single vector that goes directly from the starting position to the ending position, representing the combination of many vectors (sets of instructions) in between

Think about...

Step D
By following all the instructions, you would have traveled a total of 6.5 km. How far from point A is point D? How do you know that it must be less than 6.5 km?

Check Your Understanding

23. Suppose you followed this orienteering path:
2.0 km at a bearing of 270°, then
2.0 km at a bearing of 225°

Find the resultant vector. Use trigonometry to find how far south and west you went.

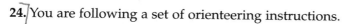

24. You are following a set of orienteering instructions.

Bearing	Distance	
0°	2.0 km	THEN
45°	1.0 km	THEN
20°	3.0 km	

(a) How far are you from the start?

(b) Estimate the bearing that would have taken you directly from start to finish.

25. Suppose the first path in Investigation 5 had been at a bearing of 45° instead of 90°.

(a) How far are you from the original starting point?

(b) Under what bearing would you have to travel to get from finish to start?

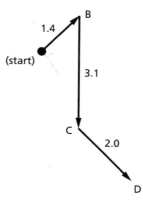

26. Find this map of the Maritimes, follow the instructions to travel from Moncton in New Brunswick. Find out where you finished and the total distance east that you have traveled.

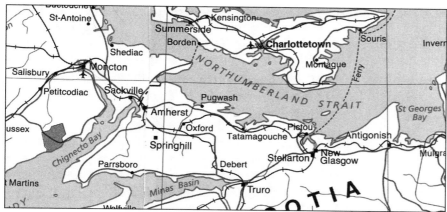

5.5 Applications of Trigonometry 247

Heading (Clockwise from North)	Distance
129°	136.9 km
76°	103.4 km
67°	144.5 km

27. You are pulling a wagon. The handle is at an angle of 30° from horizontal. Your pulling force is 20.0 newtons. Find the vertical and horizontal forces, in newtons, on the wagon.

— Note —
A newton is a measure of force.

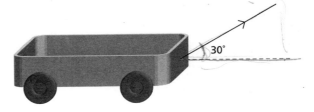

28. You are walking up a hill at a pace of about 4.0 km/h. The hill makes an angle of 10° with the horizontal.
    (a) If you have gone 10.0 km, how far have you moved horizontally? vertically?
    (b) What were your horizontal and vertical speed components?

29. What measurements could you use to find the width of the river? Write a description of the measurements you would need to find the width of the river and how you would use them.

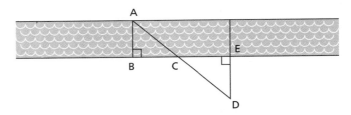

CHALLENGE yourself

You have agreed to meet a friend at a particular place in the Fundy National Park. You are 110 m apart. He must walk at an angle of 40° to reach the spot. You must walk at an angle of 60° to reach the spot. How far do you each have to walk?

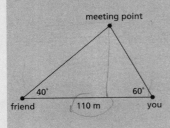

Chapter Project

Planning a Roof

Conclusion

(a) Complete the roof by calculating the amount of plywood you would need to build the roof of your choice. You can use the triangular bases from any of the plans you have already created or you can create a new base. You must show how to use trigonometry to calculate some of the areas.

(b) Which roof plan do you think is the best? Why?

PUTTING IT TOGETHER

CASE STUDY 1

Find the height of the tallest tree near to your school, taking as few measurements as possible.

Use one method involving similar triangles and no trigonometric ratios. Use another method involving trigonometric ratios.

CASE STUDY 2

Create a collage to show your understanding of the basic ideas in this chapter. It must include:

- a section with an original problem that is most efficiently solved using similar triangles;
- a section showing the meaning of the Pythagorean theorem visually;
- a section showing how $\sqrt{58}$ relates visually to 58;
- a section showing when the cosine of an angle is greater than the sine and why; and
- a section with an original problem that can be solved using tangents.

EXTENSION 1

(a) Draw semicircles on the three sides of a right triangle. Determine the area of each. What do you notice?

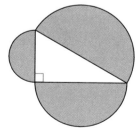

(b) Build a different shape on the hypotenuse of your right triangle. The hypotenuse must form one side of the shape. Build similar shapes on the other legs. Find all the areas. What do you observe?

EXTENSION 2

You can build your own device for measuring angles of elevation and use it to find heights. You will need a protractor, a drinking straw, some string, and a weight. To make the device, follow these steps:

1. Tape a straw along the diameter of the semicircle forming a protractor.
2. Place a tiny hole in the centre of the straw and thread the string through the hole.
3. Tie a weight to the other end of the string.
4. If you wish, mount the device on a plywood backing.

To use the device, notice that when you look at something directly at eye level through the straw, the string hangs over the 90° mark on the protractor. This represents 0°.

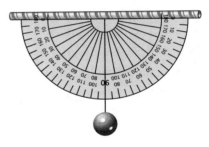

Figure out how the device could be used to measure heights and explain how and why it works.

EXTENSION 3

Kevin and Larry agreed to meet at a certain point in the woods. Kevin must walk 2.828 km at a bearing of 45° to get there. Larry must walk 3.162 km at a bearing of 198.435° to get there.

(a) How far apart are Kevin and Larry at the start?
(b) If Kevin was at the point (1,2) at the start, at what coordinates would Larry have been at the start?
(c) At what coordinates would Jamie be if he would have to walk 3.0 km at a bearing of 315° to meet Kevin and Larry at where they are meeting?
(d) Create a similar problem using different measurements. Create also a solution key to check work by other students.

Once you find the solutions to (b) and (c), verify that they are correct, using a method different from the one used to find the solutions.

REVIEW

Key Terms

	page
adjacent side	214
angle of elevation	244
bearing	217
congruent	213
cosine	236
dilatation	213
hypotenuse	214
opposite side	214
principal square root	226
Pythagorean theorem	220
Pythagorean triple	224
radical	226
resultant vector	246
similar	213
sine	236
tangent	236
trigonometric ratios	236
vector	217

You Will Be Expected To

- use similar right triangles to find the lengths in one triangle given appropriate information about the similar triangle.
- recognize the importance of the Pythagorean theorem both in finding the missing side or angle of a right triangle and determining whether a triangle is a right triangle.
- understand a proof of the Pythagorean theorem.
- describe certain square roots in terms of the lengths of sides of right triangles and in relation to other square roots.
- describe the meaning of the sine, cosine, and tangent functions and apply them to solve right triangles.
- recognize the relationship between sine and cosine of the two non-right angles in a right triangle.
- recognize the types of problems that can be solved using trigonometry and set them up appropriately for solution.
- recognize the difference between a proof and an indication that something might be true.

Summary of Key Concepts

5.1 Ratios Based on Right Triangles

When two triangles are similar, proportions can be used to measure indirectly.

Example
To find the width of the river, pace 100 m down the bank of the river. Sight a point directly east from your starting point on the other bank. Measure the angle of view x.

Now make a triangle which is similar to $\triangle$EWS, which is 10 cm high and has an angle of x degrees.
Measure XF in centimetres.

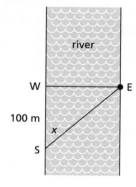

$XF \div 10 = WE \div 100$

$WE = 10\ XF$

The width of the river will be WE in metres.

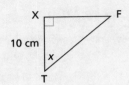

5.2 The Pythagorean Theorem

The Pythagorean theorem states that in any right triangle, the sum of the squares of the side lengths is the square of the hypotenuse length. This theorem can be interpreted geometrically to mean that the two squares built on the legs together fit exactly into the square on the hypotenuse. You should be familiar with at least one proof of the theorem.

Example

You want to know the distance across a lake.

Walk due south from the east end of the lake until you are in a line of sight with the west end.

Measure ES and WS. Find EW by using the Pythagorean theorem.

$EW^2 = SW^2 - ES^2$

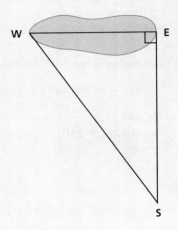

5.3 Square Roots and Their Properties

You should recognize that certain square roots can be represented as sides of right triangles.

Example

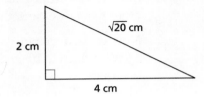

Since $20 = 2^2 + 4^2$, then $\sqrt{20}$ is the hypotenuse length of a 2-4-$\sqrt{20}$ right triangle.

In addition, the triangle above is similar to the right triangle with side lengths of 1-2-$\sqrt{5}$. It is twice as high. Therefore, $\sqrt{20}$ can be simplified to be $2\sqrt{5}$.

Another way to simplify $\sqrt{20}$ is to note that

$$\sqrt{20} = \sqrt{4 \times 5} = \sqrt{4} \times \sqrt{5} = 2\sqrt{5}$$

5.4 Defining Trigonometric Ratios

Parallel lines have equal slopes. Perpendicular lines have slopes that are negative reciprocals. In the triangle,

$$\sin B = \frac{\text{opposite}}{\text{hypotenuse}} = \frac{AC}{AB}$$

$$\cos B = \frac{\text{adjacent}}{\text{hypotenuse}} = \frac{BC}{AB}$$

$$\tan B = \frac{\text{opposite}}{\text{adjacent}} = \frac{AC}{BC}$$

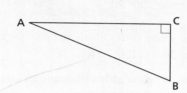

Example

If you know one of the non-right angles and any side, you can find the other sides and other angles.

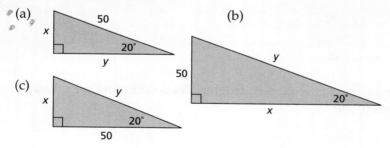

In figure (a), you know that $\sin 20° = \frac{x}{50}$. Since $\sin 20° = 0.34$, $x = 17$. Since $\cos 20° = \frac{y}{50}$ and $\cos 20° = 0.94$, $y = 47$. You could also have calculated y using the Pythagorean theorem, once you knew x.

In figure (b), you know that $\sin 20° = \frac{50}{y}$. Since $\sin 20° = 0.34$, $y = 150$. Since $\tan 20° = \frac{50}{x}$ and $\tan 20° = 0.37$, $x = 140$.

In figure (c), you know that $\cos 20° = \frac{50}{y}$. Since $\cos 20° = 0.94$, $y = 53$. To find x, you could either use the fact that $\tan 20° = \frac{x}{50}$, so that $x = 18$ or you could use the fact that $\sin 20° = \frac{x}{y}$.

Sometimes, the situation is slightly more complicated. For example, you might have a non-right triangle and know the angles and one side length. With this information, you can find the other side lengths.

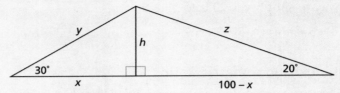

$\tan 30° = \dfrac{h}{x}$, so $h = x \tan 30°$.

However, $\tan 20° = \dfrac{h}{100 - x}$, so $h = (100 - x) \tan 20°$.

Setting the two values for h to be equal, you conclude that
$x \tan 30° = (100 - x)\tan 20°$.

Using a calculator or a table of trigonometric ratios to substitute values for $\tan 20°$ and $\tan 30°$, you can find that $x = 39$.

You can now find h, since $h = x \tan 30°$, $h = 22$. The other two unknown side lengths can be found using the Pythgorean theorem or trigonometric functions.

$\sin 30° = \dfrac{h}{x}$, so $y = 45$.

$\sin 20° = \dfrac{h}{z}$, so $z = 65$.

In other cases, you know the side lengths and need to find the missing angle. For example,

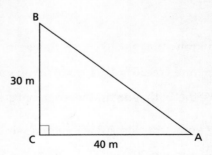

You know that the tangent of angle A is $\dfrac{30}{40}$. To find the measure of angle A, you can try different values for A on the calculator until you get the value 0.75. In this case the angle would be about 37°.

PRACTICE

5.1 Ratios Based on Right Triangles

1. John built a bike jumping ramp. It has a rise of 1 cm for every 10 cm of run. If the rise is 22 cm, how long is the ramp along the floor?

2. To find the height of a building, Amy measured the shadow of her 30 cm ruler to be 55 cm right in front of a building that had a shadow of 27 m at the same time. How tall was the building?

5.2 The Pythagorean Theorem

3. Draw two right triangles and verify that their side lengths are consistent with the Pythagorean theorem.

4. Describe a proof of the Pythagorean theorem.

5. Describe a way to test if a triangle is a right triangle without using a protractor. Is it possible by just knowing the side lengths?

5.3 Square Roots and Their Properties

6. Simplify.
   (a) $\sqrt{80}$
   (b) $\sqrt{360}$
   (c) $\sqrt{3600}$
   (d) $5\sqrt{2} + 8\sqrt{8}$
   (e) $6\sqrt{34} + 2\sqrt{136}$

7. Draw a right triangle with a hypotenuse of $\sqrt{85}$ units. Then draw one with a hypotenuse of $\sqrt{340}$ units. Why is the second one easy to do after the first one is drawn?

8. How is $\sqrt{2500}$ related to $\sqrt{25}$? Why?

9. Describe length $\overline{AX}$ as a square root in two different ways.

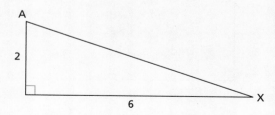

5.4 Defining Trigonometric Ratios

10. When can sin X = cos X? Why? When would sin X > cos X?

11. Kevin found the sine of an angle on his calculator and read it off as 1.1917536.
    (a) How do you know that he is wrong?
    (b) What trigonometric function should that number be describing?
    (c) Use the trigonometric function from (b) to find the size of angle X.

12. Why does the value of sin 30° not change if the triangle in which the 30° is found is enlarged?

13. (a) Draw a right triangle in which sin X is about 0.64. Predict the angle size before you check with your calculator or a trigonometric table.
    (b) Draw a right triangle in which cos X is about 0.64. Why is this easy after you have completed (a)?
    (c) Draw a right triangle in which tan X is very close to 0.5. Predict the angle size before you check with your calculator or a trigonometric table.

14. The sine button on your calculator is not working. How can you use the diagram to give you another idea for finding the value of sin 66°?

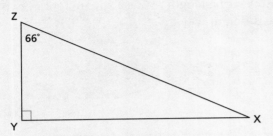

5.5 Applications of Trigonometry

15. You are measuring trees in the forest. You are only allowed to cut down trees between 20 m and 30 m in height. You see the top of one tree at an angle of elevation of 22° from a ground level position 40 m from the base of the tree. Should you cut it down? Explain.

16. You need to reach a window 3.8 m from the ground. The ladder tilt is 70°. How long is the ladder?

17. Find the area of this trapezoidal piece of land.

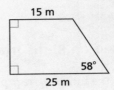

18. Create a word problem which could be solved by finding tan 35°.

19. Describe the area of an isosceles triangle in terms of the tangent of the base angle and the length of the base. Try out your formula with several triangles.

20. Suppose you can get from your front door to your friend's front door by walking east for 320 m and then walking at a bearing of 45° for 220 m. How far is your friend's front door from yours?

Chapter Six
The Geometry of Packaging

Some snack foods are packaged in airtight bags to keep them fresh. Can you think of any other types of packaging? Many food products are packaged in rectangular boxes. What other package shapes are common in grocery stores? Why do you think different shapes are used?

Packaging engineers and designers must develop packaging that protects the product while meeting the needs of consumers. When containers are designed and constructed, the weight, strength, flexibility, cost, and means of disposal need to be considered. The designer also has to think about the tools and machines that are needed to cut and assemble the container.

After successfully completing this chapter, you will be expected to:

1. Solve problems using the formulas for the surface area and volume of prisms, pyramids, cylinders, and cones.

2. Understand the properties and symmetry of two-dimensional and three-dimensional shapes.

3. Solve problems using the relationships between perimeter and area, and surface area and volume.

4. Understand the concepts of area of two-dimensional shapes and surface area and volume of three-dimensional shapes.

5. Apply relationships between the perimeters and areas of similar two-dimensional shapes, and between the surface areas and volumes of similar three-dimensional shapes.

6. Apply altitude, median, angle bisectors, and perpendicular bisectors.

6.1 Examining Factors in Container Design

Atlantic Packaging is a company that designs and manufactures containers. Their containers vary from rectangular prism boxes and cylindrical cans to triangular prism cardboard tubes.

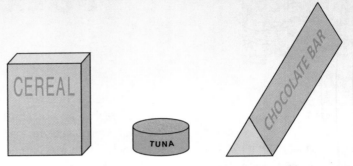

In many cases, the shape and dimensions (width, depth, and height) of the container are determined by the shape and dimensions of the product, for example, a pair of shoes or a television. Now, you will be looking at containers for products that do not have a specific shape, for example, breakfast cereal and cooking oil. The shape and dimensions of containers for these types of products depend on a variety of factors.

This section will focus on how designers use geometry when designing a container. You will investigate how to design the most economical container possible.

One of the important factors to consider when designing a container for a product is volume. To keep costs down, the container should have a volume that is as close as possible to the volume of the product.

volume – the amount of space an object takes up. It is measured in cubic units such as cubic centimetres (cm³) or cubic metres (m³).

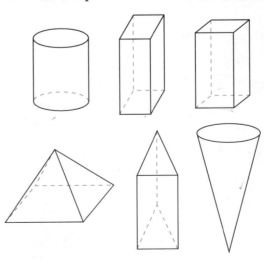

Think about...

Equal Volume

All these containers have the same volume.

- Identify each shape.
- List advantages and disadvantages of the different container shapes.

258   Chapter 6 *The Geometry of Packaging*

Volume Formulas

The cross-section of prisms and cylinders is the same shape over the entire height. For shapes like these, the volume is the cross-sectional area × height, and can be determined by using the formula $V = Ah$, where h is the height and A is the area of the base.

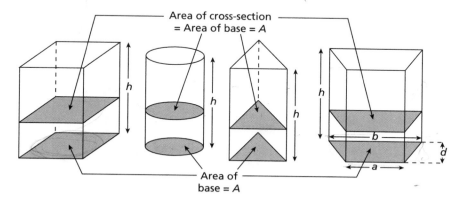

The volume of a pyramid or cone can be determined by the formula $V = \frac{1}{3}Ah$.

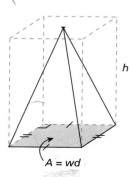
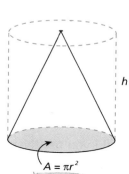
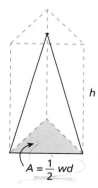

The volume of a sphere can be found using the formula $V = \frac{4}{3}\pi r^3$.

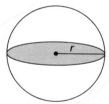

r is the radius of the sphere

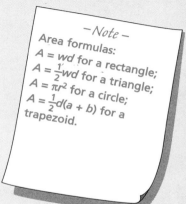

– Note –
Area formulas:
$A = wd$ for a rectangle;
$A = \frac{1}{2}wd$ for a triangle;
$A = \pi r^2$ for a circle;
$A = \frac{1}{2}d(a + b)$ for a trapezoid.

– Note –
Water poured from a cone into a cylinder with the same height and radius fills the cylinder $\frac{1}{3}$ full.

6.1 Examining Factors in Container Design

Focus Questions

1. (a) A large block of cheese has dimensions as shown. What is its volume?

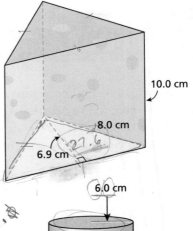

(b) A juice can has the dimensions shown. What is its volume?

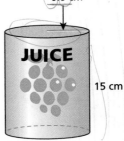

(c) A box of French fries has dimensions as shown. Determine its volume.

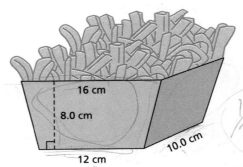

2. (a) For which shapes could you use the volume formula $V = Ah$?

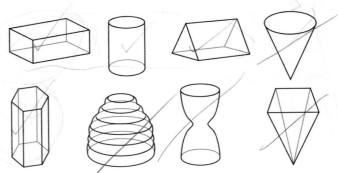

(b) Explain in words and with diagrams why the volume formula may or may not work for each shape.

— Note —
Don't forget to use the significant digit rules! When multiplying or dividing, your results should be recorded with the same number of significant digits as the least precise measurement in the question.

Think about...

Volume
List five factors that might determine the volume of a product to be packaged.

3. Determine how much space there is in the pup tent.

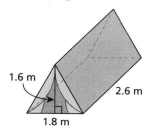

4. A slushy container looks something like the diagram. Determine the volume of the slushy the container would hold.

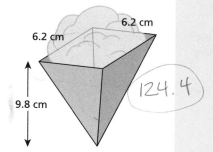

124.4

5. (a) Determine the volume of concrete needed to build each pillar.

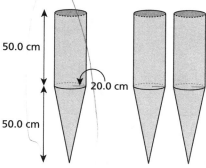

(b) How much hay could the hayloft on the right contain? The hayloft consists of the top two parts of the barn.

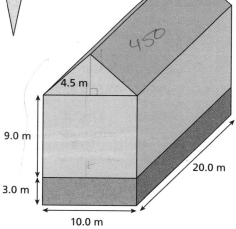

450

(c) How much will the Jell-O mould hold? The Jell-O mould includes only the top part of the cone.

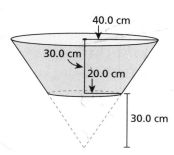

— Note —
Remember that 3.14 is only an approximation of π. π is not equal to 3.14. Most calculators have a π button that you can use for your calculations.

6.1 Examining Factors in Container Design 261

Think about...

Cone and Pyramid Containers

List five examples each of cone-shaped and pyramid-shaped packaging containers.

— Note —
Question 7
1 L = 1000 mL
= 1000 cm³

capacity — the amount of space inside a container. It is usually measured in millilitres (mL) or litres (L) (1 mL = 1 cm³).

(d) Adam and Erin are filling their aunt's old mailbox completely with gravel to make a doorstop for a barn door. How much gravel will they need? The top of the mailbox is a half cylinder.

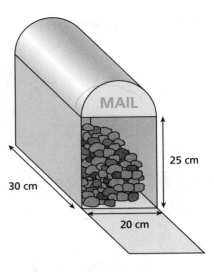

6. Explain, in writing, how you could find the half-full level of each of these containers.

7. (a) A rectangular prism box has a capacity of 3 L.

One possible set of dimensions is given in the following chart. Copy and complete the chart in your notes by adding three other sets of dimensions that would produce a container with the same **capacity**. Try not to use the same numbers in several sets.

Width	Depth	Height	Volume/Capacity
15 cm	10 cm	20 cm	3000 cm³ = 3 L

(b) Select the set of dimensions that you feel is most likely for a cereal box. Explain your choice.

8. Draw a similar chart for cylindrical containers designed to hold 1 L of liquid. Since the cylinder formula uses π, it may be difficult to find dimensions that produce capacities of exactly 1 L. Try to get as close to 1 L as you can.

The following chart may help you get started.

Radius of Base	Height	Volume/Capacity
4.6 cm	15 cm	997 cm³ ≈ 1 L

Chapter Project

Designing Containers

Your project is to design a container that a manufacturer could use for packaging tennis balls.

Try to build a container that requires the least packaging material. Once the material is chosen, the cost of the container depends on the amount of material needed to construct it.

Your project consists of three basic parts:

A. Detailed, well-organized solutions to all the questions labeled Project Connection in each section of this chapter.

B. The final design for your container and clear explanations for your choice of material, shape, and dimensions, as well as all sketches and accurate or scale drawings.

C. A full-sized model of your container. (The material you use to build your model (prototype) may be different from the material you would use to manufacture the container.) Your container design must be well thought out and carefully planned and constructed with clear explanations for your choice.

— Note —
Here are some ideas to get you started.
- Choose the number of tennis balls that you will be packaging.
- Choose the material that you will use to make your container.
- Think about the mathematics you will need to complete your project.

Project Connection

Suppose you have an orange that is almost perfectly round. The volume of the orange includes the skin and everything inside.

1. How could you find the volume of the orange in at least two different ways? Estimate the volume of the orange.

2. Is it possible to determine the radius of the orange without cutting it? Explain how and then do it.

3. Imagine packing a round object into a cylinder that has width, depth, and height equal to the diameter of the object.
   (a) Use a piece of paper to completely enclose an orange.
   (b) Measure to find the volume of the cylinder formed by the paper.
   (c) Estimate the volume of the orange and compare it to the volume of the cylinder.

4. The radius of a field hockey ball is about 3.7 cm. Find its volume.

5. Why is volume measured in cubic units?

6. If you had to fit a sphere of radius r exactly inside a cylinder, how would the volumes of the sphere and the cylinder compare?

7. Determine the radius of a bubble with a volume of 20.0 cm^3.

6.2 Regular Polygons

Atlantic Packaging is designing a new line of prism-shaped boxes. Their cross-sectional shapes—regular pentagons, hexagons, heptagons, and octagons—will make them novelty items suitable for gift boxes.

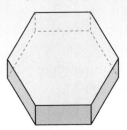 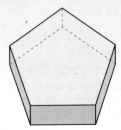

In this section, you will develop a method for finding the area of any regular polygon. You will then be able to use it to find the volume of any regular polygonal prism.

FOCUS B Regular Polygons

What is a regular polygon?

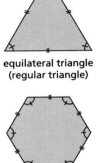

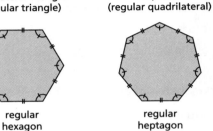

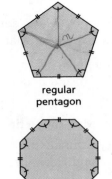

equilateral triangle
(regular triangle)

square
(regular quadrilateral)

regular pentagon

regular hexagon

regular heptagon

regular octagon

A regular polygon is a two-dimensional closed figure with three or more sides. All sides are equal and all angles are equal. The polygons above are all regular polygons.

264   Chapter 6 *The Geometry of Packaging*

Containers often exhibit symmetry. Designers want them to be visually appealing and functional. Regular polygonal prisms have rotational symmetry equal to the number of sides of the polygon. This makes it easier to design a lid and packaging for the container.

What are some special characteristics of regular polygons?

Regular polygons have both **line** and **rotational symmetry**.

Explain how you could use each type of symmetry to locate the centre of a regular polygon.

(a) line symmetry (b) rotational symmetry

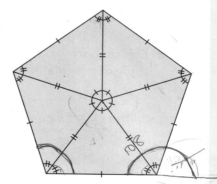

line symmetry – the property of a shape where it can be folded in half and the halves match. A square has 4 lines of symmetry.

rotational symmetry – the property of a shape where a tracing of it is turned around its centre and the shapes match at least once before the tracing completes a full rotation

An equilateral triangle has an **order of rotational symmetry** of 3 because it fits 3 ways in one full turn.

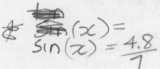

Focus Questions

1. Without measuring, determine the unknown angles in each regular polygon.

   (a) 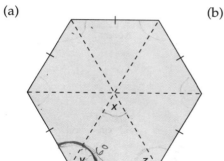   (b)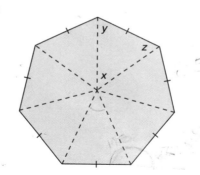

   (c) Describe to your friend how to find the angles at the vertices of a regular octagon.
   (d) Find the sum of the **interior angles** of any pentagon.
   (e) Using only one angle at each vertex, find the sum of the **exterior angles** of a pentagon.
   (f) Using only one angle at each vertex, find the sum of the exterior angles of the polygons in parts (a) and (b). What do you notice? Make a conclusion based on your findings.

interior angle – the angle at each vertex of a polygon. There are six interior angles in this hexagon.

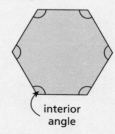

interior angle

exterior angle – the angle formed with an adjacent side of a polygon when one side of the polygon is extended. There are 12 exterior angles in this hexagon.

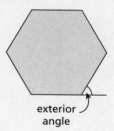

exterior angle

2. (a) Explain how you could determine the measure of an exterior angle of a regular **dodecagon** (12-sided polygon). What is the measure?

(b) How could you use the answer to part (a) to determine the measure of an interior angle of a dodecagon?

(c) Explain how you could determine the measure of an exterior angle of a regular n-sided polygon.

(d) Explain how you could determine the measure of an interior angle of a regular n-gon.

Investigation 1
Determining the Area of Regular Polygons

Purpose
To find and compare the perimeter and area of six regular polygons.

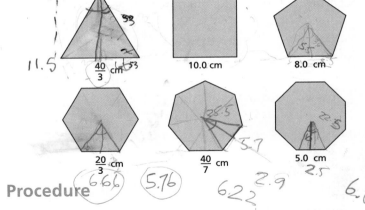

Think about...

Naming Polygons

What is a more familiar name for:
- a regular triangle?
- a regular quadrilateral?

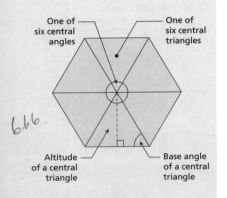

One of six central angles
One of six central triangles
Altitude of a central triangle
Base angle of a central triangle

Procedure

A. Copy and complete the fourth column of the table. The fifth column will be completed in Step B.

Number of Sides	Length of Each Side	Perimeter	Area of Polygon	Volume of Prism
3	$\frac{40}{3}$ cm	40 cm	76.6	153.2
4	10.0 cm	40 cm	100	2000
5	8.0 cm	40 cm	110	2200
6	$\frac{20}{3}$ cm	40 cm	114.3 115	2286.1 2300
7	$\frac{40}{7}$ cm	40 cm	118.9 121	2378 2436
8	5.0 cm	40 cm	120	2400

266   Chapter 6   *The Geometry of Packaging*

B. Suppose each polygon is the base of a 20-cm high prism. Determine the volume of each prism. Complete the fifth column of the table. What do you notice?

Investigation Questions

3. Explain what you did to find the area of each polygon in Investigation 1.

4. (a) What do you notice about the perimeter of each polygon?
   (b) Discuss what happened to the area of the polygons with fixed perimeters as the number of sides increased.
   (c) Describe, in writing, what happens to the area of the regular polygon as the number of sides increases and the perimeter remains fixed.
   (d) If all the polygons had the same area, would the perimeter increase or decrease as the number of sides of the polygon increased?
   (e) How can you find the area of any regular polygon using the perimeter and the **apothem**?

apothem – the perpendicular distance from the centre of any regular polygon to a side

Check Your Understanding

5. Explain how you could determine the areas of an equilateral triangle and a square without dividing them into triangles with vertices at the centre.

6. Each regular polygon has a perimeter of 100 cm. Find the areas.

	Perimeter (cm)	Area (cm²)
(a) Equilateral Triangle	100.0	481
(b) Square	100.0	625
(c) Regular Pentagon	100.0	688
(d) Regular Hexagon	100.0	721
(e) Regular Heptagon	100.0	748
(f) Regular Octagon	100.0	754

7. (a) Make a prediction about the area of a regular **nonagon** (9-sided polygon) with a perimeter of 100 cm. Calculate to check.
   (b) Sketch a graph of Number of Sides versus Area. Check your prediction against the graph.

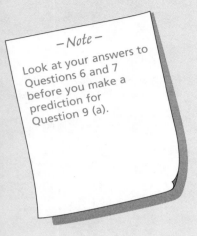

— Note —
Look at your answers to Questions 6 and 7 before you make a prediction for Question 9 (a).

8. (a) For regular polygons with a fixed perimeter, describe the relationship between the number of sides and the area.
   (b) For regular polygons with a fixed area, describe the relationship between the number of sides and the perimeter.

9. A circle has a circumference of 100.0 cm.
   (a) Predict its area. Explain your prediction.
   (b) Calculate to check. Explain what you did to find the area of the circle.
   (c) Why do you think the circle is called the limiting shape for a regular polygon?

10. Use a piece of 22-cm by 28-cm paper.
    (a) Construct an equilateral triangle prism tube. What is its capacity?

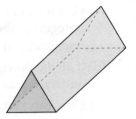

   (b) Use the same piece of paper to create a regular hexagonal tube with the same height.

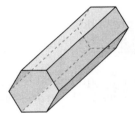

      (i) Predict whether its capacity will be greater or less than the capacity of the triangular tube. Explain your prediction.
      (ii) Check by calculating its capacity.
   (c) Which tube is more economical? Explain.
   (d) Research the best shape for an eavestrough. Explain why you think it is the best.

11. What is the volume of a regular pentagonal prism if it has a height of 15 cm and a base perimeter of 25 cm?

Think about…
Economical Containers
What factors might make one container more economical than another?

12. Susan wants to completely fill the tin at the right with fudge as a present for her grandmother. How much fudge should she make?

13. What is the capacity of a pyramid-shaped box that is 20 cm tall and has a regular hexagonal base with side lengths of 15 cm?

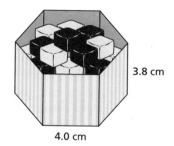

3.8 cm

4.0 cm

14. The regular nonagon has the greatest symmetry of all regular polygons. How would you prove or disprove this to the class?

15. Each of these lids has the same perimeter and fits on a 1-L gift box.

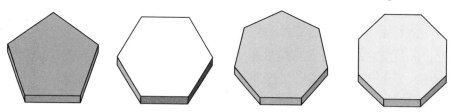

(a) What do you know about how the areas of the lids compare?
(b) Order the boxes from least to greatest height. Justify the order.

16. Write to a classmate explaining what happens to the area of a polygon as the number of sides increases (assuming that the perimeter is held constant).

17. (a) Describe, in writing, a method for finding the area of any regular polygon if you know its perimeter and number of sides.
    (b) Express your written description as a formula.

Project Connection

1. Determine the volume of each container. The containers are designed to contain one sphere with a diameter of 8.0 cm.

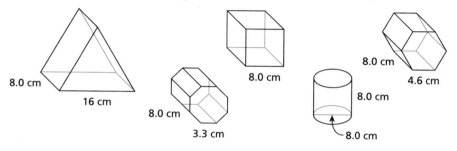

2. Determine the percent of each container filled by the sphere.

3. Which container minimizes the amount of empty space?

Did You Know?

Bees' honeycombs are made of regular hexagonal prisms. This shape allows the bees to store the most honey using the least amount of beeswax.

Did You Know?

The Canadian loonie is a regular **hendecagonal** (11-gonal) prism. Take the appropriate measurements to determine its volume.

Think about...

Container Design
Volume and capacity are important factors in the design of a container. Why do you think surface area needs to be considered in the design of an economical container?

6.3 Surface Area

A large part of the cost of a container is determined by the amount of material used in its construction. The less material needed to contain a product, the more economical the container. This section focuses on calculating the surface area of containers in order to estimate how much material is required to make them.

Focus C: Surface Area

Think about...

Cost Factors

Surface area can be used to estimate the cost of constructing a container.

Discuss other factors that might affect the cost.

To find the surface area of a three-dimensional object with polygonal faces, add the areas of the faces.

For example, for a triangular prism, find the sum of the areas of the two triangular faces, and the three rectangular faces.

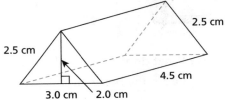

Surface area = $(2 \times \frac{1}{2} \times 3.0 \times 2.0) + (4.5 \times 2.5) + (4.5 \times 2.5) + (4.5 \times 3.0)$

$= 42$

The surface area of the prism is 42 cm².

Some objects do not have polygonal faces, such as cones, spheres, and cylinders. Thus, their surface areas cannot be found by finding the areas of polygons.

The surface area of a cone is given by $\pi r^2 + \pi rs$, where r is the radius of the circular base, and s is the slant height of the cone.

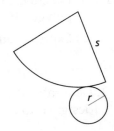

lateral surface – the curved surface of a cylinder that unfolds to a rectangle in net form. Its dimensions are the circumference of the base (c) and the height of the cylinder (h).

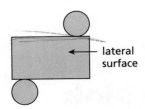

The surface area of a sphere is given by $4\pi r^2$ where r is the radius of the sphere.

lateral face – a side face of a prism. All the lateral faces, not including the two bases, make up the **lateral surface** of a prism.

The surface area of a cylinder is given by $2\pi r^2 + 2\pi rh$, where r is the radius of the circular base, and h is the height of the cylinder.

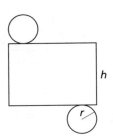

Focus Questions

1. Find the surface area of each object.

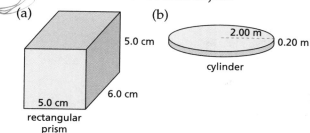
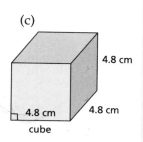

(a) rectangular prism: 5.0 cm, 5.0 cm, 6.0 cm
(b) cylinder: 2.00 m, 0.20 m
(c) cube: 4.8 cm, 4.8 cm, 4.8 cm

2. Find each surface area.

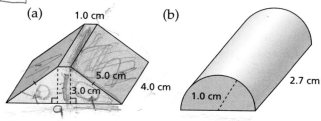
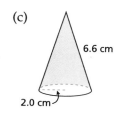

(a) 1.0 cm, 5.0 cm, 3.0 cm, 4.0 cm
(b) 2.7 cm, 1.0 cm
(c) 6.6 cm, 2.0 cm

Did You Know?

Grace Chisholm Young (1868–1944) was the first woman to receive an official doctorate in mathematics in Germany. In her *First Book of Geometry* she used nets to help students understand how to work with three-dimensional figures.

Investigation 2

Surface Area and Container Design

Purpose

In this Investigation you will calculate and compare the volumes and surface areas of different-shaped containers.

Procedure

A. Calculate and compare the volumes and surface areas of these square-based prisms.

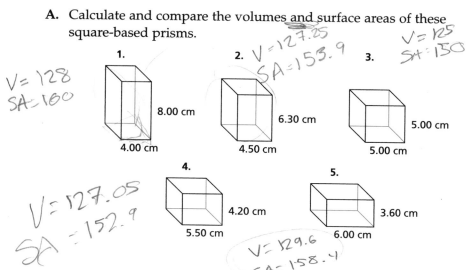

1. 8.00 cm, 4.00 cm V = 128 SA = 160
2. 6.30 cm, 4.50 cm V = 127.25 SA = 153.9
3. 5.00 cm, 5.00 cm V = 125 SA = 150
4. 5.50 cm, 4.20 cm V = 127.05 SA = 152.9
5. 6.00 cm, 3.60 cm V = 129.6 SA = 158.4

B. Calculate and compare the volumes and surface areas of these cylinders. The diameter and height are shown.

1. 10.2 cm, 4.00 cm — V=128.1, S=153.3
2. 8.00 cm, 4.50 cm — V=144.9, S=127.2
3. 6.50 cm, 5.00 cm — S=141.4, V=127.6
4. 5.40 cm, 5.50 cm — V=85.9, S=108.1
5. 4.50 cm, 6.00 cm — V=127.2, S=141.4

C. Describe any patterns you noticed while comparing the results of your calculations for Steps A and B.

Check Your Understanding

3. (a) Suppose the cost of metal to make these containers is 2¢/cm². Copy and complete the table.

Dimensions of Container	Volume (cm³)	Surface Area (cm²)	Cost ($)
i. 12.00 cm wide × 20.00 cm deep × 10.00 cm tall	2400.0 cm³	1040 cm²	$20.80
ii. 8.00 cm wide × 15.00 cm deep × 20.00 cm tall	2400 cm³	1520 cm²	$30.40
iii. 6.00 cm wide × 16.00 cm deep × 25.00 cm tall	2400 cm³	1900 cm²	$38.00
iv. 15.00 cm wide × 16.00 cm deep × 10.00 cm tall	2400 cm³	940 cm²	$18.00

(b) Which container is the least expensive to make? Explain.

4. (a) Determine the surface area of this wedge of wood.

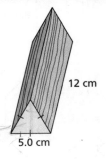

12 cm, 5.0 cm, 4.3

2△ = 21.5

272   Chapter 6 *The Geometry of Packaging*

(b) Determine the surface area of this pentagonal play tunnel.

5. (a) Suppose each of the following 1-L containers were made of the same material. This material costs 1¢/cm². Which one would cost the least to manufacture? Explain.

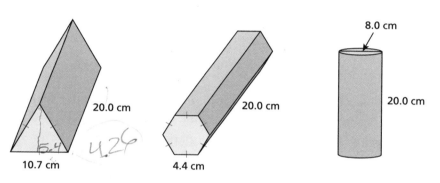

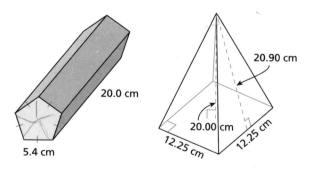

(b) Describe how to change the dimensions, while retaining the shape of each container, so that each container would hold the following:

(i) 8 L   (ii) 0.125 L

6. Suppose each container in Question 5 is open-topped. Which is the most economical container now? Explain.

CHALLENGE
yourself

For the containers in Question 5, graph the surface area versus the cost. What do you notice about the graph?

— Note —
In the pyramid in Question 5, the pyramid height is 20 cm and the slant height is 20.90 cm. Which is used to find the volume? Which is used to find the surface area?

6.3 Surface Area   273

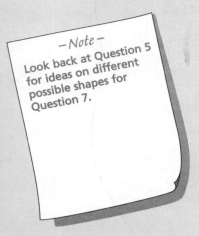

— Note —
Look back at Question 5 for ideas on different possible shapes for Question 7.

7. Design two containers that have approximately the same volume but different shapes.
   (a) Predict which has the greater surface area. Explain your prediction.
   (b) Calculate the surface areas to check.
   (c) See if you can find examples at home of different-shaped containers with about the same volume.

Project Connection

The surface area of an orange is the area of the outside skin of the orange.

1. Estimate the surface area of the orange in at least two different ways. You can get a good estimate of the orange's surface area by peeling it and placing the peel on graph paper. Then count the squares. If you use small-grid graph paper, your estimate should be reasonably accurate.

   Find another way.

2. Suppose you were packing a round object into a cylinder with width, depth, and height equal to the diameter of the object.
   (a) Using an orange and a piece of paper, construct a cylinder to enclose the orange.
   (b) Calculate the surface area of the orange and the cylinder (not including its ends).
      How does the surface area of the cylinder compare to the surface area of the orange that fits inside it?
      Use the formula to get exact values and compare them again.

3. The radius of a field hockey ball is about 3.7 cm. Determine its surface area.

4. Why is surface area measured in square units?

5. If you had to fit a sphere of radius r exactly inside a cylinder, how would the surface areas of the sphere and the cylinder compare?

6. The surface area of a basketball is about 1810 cm². Describe how you could determine its radius using this information.

7. Determine the radius of a bubble with a surface area of 20.0 cm².

8. The outside of a baseball is made of two congruent pieces of stitched leather. Find the area of one piece if the diameter of a baseball is about 7.5 cm.

6.4 Economy of Design

In Section 6.3, you discovered that some shapes and dimensions of containers are more economical than others because they hold the same amount using less material. The focus of this section is on developing a method to rank containers according to their economy rate and then determining the most economical shape for any volume of container.

Investigation 3
Exploring the 1-L Milk Carton

Purpose
To compare the amount of material used to construct different 1-L cartons.

Procedure

A. Calculate the volume and surface area of a 1-L (1000-cm³) milk carton. Assume the carton has a flat top.

B. Copy and complete the table with four more possible sets of dimensions for a 1-L milk carton.

Think about...

Milk Cartons
Look closely at a 1-L milk carton. Describe how it is constructed.

rate – a number that compares two quantities with different units

economy rate of a shape – a rate that compares the volume of a container to its surface area. A high *ER* means a container is more economical. A low *ER* means a container is less economical.
$ER = \frac{V}{SA}$.

—Note—
A spreadsheet is a useful tool for creating a table of values.

Volume, V (cm³)	Width, w (cm)	Depth, d (cm)	Height, h (cm)	Surface Area, SA (cm²)	Economy Rate, ER
1.00×10^3	7.07	7.07	20.0	665	1.5
1.00×10^3	10	10	10	600	1.66
1.00×10^3	5	5	40	850	1.18
1.00×10^3	9	9	12.3	604.8	1.65
1.00×10^3					

Think about...

Decimal Forms

Why is the decimal form of a rate better for comparing economy rates? Use an example in your explanation.

Investigation Questions

1. Two containers with the same volume have different economy rates—one has a rate of 0.3 and the other of 0.5. Which container is more economical? Explain.

2. How does the shape of the most economical container you found compare to the shape of the actual milk carton?

3. Why hasn't someone developed a more economical 1-L milk carton?

Investigation 4

Redesigning the 1-L Milk Carton

Purpose

To find the most economical dimensions for a 1-L carton with a height of 20 cm and a flat top.

Procedure

A. Copy and complete the table.

Volume, V (cm³)	Height, h (cm)	Width, w (cm)	Depth, d (cm)	Perimeter of Base (cm)	Area of Base, A (cm²)	Surface Area, SA (cm²)	Economy Rate, ER
1.00×10^3	20.00	4.00	12.5	33	50	760	1.32
1.00×10^3	20.00	5.00	10	30	50	700	1.43
1.00×10^3	20.00	6.00	8.3	28.6	49.8	671	1.49
1.00×10^3	20.00	7.00	7.1	28.3	50	663	1.51
1.00×10^3	20.00	8.00	6.3	28.5	50	672	1.49
1.00×10^3	20.00	9.00					

Did You Know?

The ancient Japanese art of origami is used to make decorative boxes by paper folding and without scissors or glue.

B. What relationship do you notice in the table between the dimensions of the base and the economy rate?

C. What are the dimensions of the base of the most economical carton in the table? What other set of dimensions would be even more economical?

Investigation Questions

4. Recall from Section 6.2 that the circle is the limiting shape for a regular polygon with a fixed area.
   (a) What does this mean for the perimeter of regular polygons?
   (b) Predict what shape of a 20-cm-tall prism container with a volume of 1 L would be more economical than a square-based prism. Explain your prediction.
   (c) Predict the shape of the most economical 20-cm-tall container possible with a volume of 1 L.
   Explain your prediction.
   (d) What are the dimensions of the container in part (c)?

5. Discuss and then summarize, in writing, what you know about:
   (a) the most economical rectangular-prism container for a given height and volume; and
   (b) the most economical container for a given height and volume.

Think about...

Cylinders and Square-Based Prisms

How are cylinders and square-based prisms alike? How are they different?

Check Your Understanding

6. (a) Calculate the volume of each container. What do you notice?

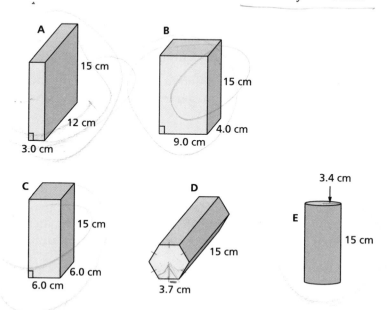

Think about...

Priorities

Why aren't all containers designed with the greatest economy rate possible?

(b) Without calculating, predict how the economy rates of the containers compare. Explain your prediction.

(c) Calculate the economy rates to check your predictions in (b).

6.4 Economy of Design   277

Investigation 5
Redesigning the 1-L Milk Carton into the Most Economical Cylinder

Purpose
To find the dimensions of the most economical 1-L cylinder.

Procedure

A. Use what you know about the most economical rectangular prism to predict the dimensions of the most economical cylinder.

B. Copy and complete the table to test your prediction.

Volume, V (cm³)	Radius, r (cm)	Diameter, d (cm)	Area of Base, A (cm²)	Height, h (cm)	Surface Area, SA (cm²)	Economy Rate, ER
1.00×10^3	4.00	8	50.3	19.9	600	1.67
1.00×10^3	4.50	9	63.6	15.7	571	1.75
1.00×10^3	5.00	10	78.5	12.7	556	1.80
1.00×10^3	5.50	11	95.0	10.5	553	1.81
1.00×10^3	6.00	12	113.1	8.8	558	1.8
1.00×10^3	6.50	13	132.7	7.5	570	1.75

C. What are the dimensions of the most economical 1-L cylinder? How does its economy rate compare to the 1-L cube?

D. Compare, in writing, the different dimensions of the most economical cylinder. Include an explanation of why it is the most economical.

SAT ERV

FOCUS D — Square Roots and Cube Roots

Did you know that 4, 9, 16, and 25 are called perfect squares because it is possible to form perfect squares with that number of unit squares?

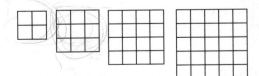

278 Chapter 6 *The Geometry of Packaging*

Similarly, 8, 27, 64, and 125 are called perfect cubes because perfect cubes are formed with that number of unit cubes.

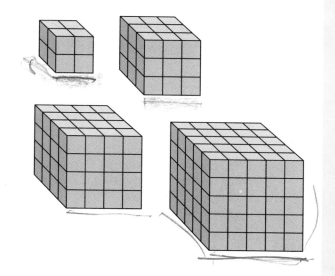

The square root of 36 is 6 since 6 is the length of a side of the square that contains 36 unit squares. The cube root of 8 is 2 since 2 is the length of an edge of the cube that contains 8 unit cubes.

If the area of a square is known, a calculator can be used to determine the side length using the square root function ($\sqrt{}$).

If the volume of a cube is known, a calculator can be used to determine the edge length using the cube root function ($\sqrt[3]{}$).

— Note —

The symbol $\sqrt[3]{x}$ is used to indicate the cube root of x. Check your calculator to see if it has a cube root key. In $\sqrt[3]{10}$, 3 is the **index** and 10 is the **radicand**.

Focus Questions

7. Explain how square roots and cube roots are alike and how they are different.

8. Calculate.
   (a) $\sqrt[3]{27}$
   (b) $\sqrt[3]{216}$
   (c) $\sqrt[3]{1000}$

9. You are given a cube with a volume of 100 cm³.
   (a) Explain how you would find, without measuring, its edge length.
   (b) Explain how you could check the accuracy of your result.
   (c) Is 100 a perfect cube? Explain why or why not.

10. The edge length of a cube is $\sqrt[3]{10}$ cm. Find its volume.

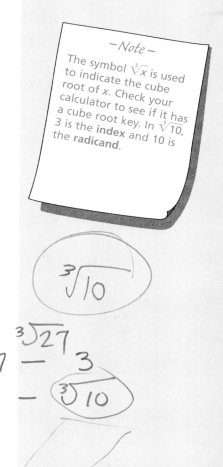

6.4 Economy of Design

Economy of Design

Example 1
You have 500 mL of cereal to package. What box dimensions would minimize the cost of the material required to make it?

Solution

Step 1: Determine the best shape for the container. In Investigation 4, you learned that the most economical rectangular prism for a given volume is a cube.

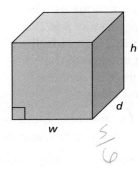

Since $V = hwd$ and, for a cube, $h = w = d$, we can represent the edge length of a cube with one variable, s. The volume of a cube can be written as

$V = s \times s \times s$ or $V = s^3$.

Step 2: Rewrite the volume formula, substituting the volume information you have.
$V = s^3$
$500 \text{ cm}^3 = s^3$

Step 3: Solve for s.
Since $s^3 = 500.0 \text{ cm}^3$
$s = \sqrt[3]{500.0 \text{ cm}^3}$
$= \_\_\_\_ \text{ cm}$

(a) What are the cube's dimensions? Use a calculator to find out.

Example 2
There is another shape of 500-mL container that would be more economical. What is it? What are its dimensions?

Solution

Step 1: Determine the shape. In Investigation 5, you learned that a cylinder with its height equal to its diameter is more economical than a cube for a given volume.

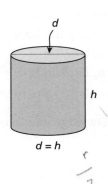

Step 2: Rewrite the volume formula, substituting what you know about the container.
$V = \pi r^2 h$
$500.0 \text{ cm}^3 = \pi r^2 h$
You also know that $h = d$ or $h = 2r$.
$500.0 \text{ cm}^3 = \pi r^2 (2r)$
$= 2\pi r^3$

Step 3: Solve for r.
$$\frac{500}{2\pi} = r^3 \quad 80 = r^3$$

$r = \underline{4.29}$

(b) What are the cylinder's dimensions? Use a calculator to find out.

Focus Questions

11. What do you know about the most economical rectangular prism that will contain each volume?
    (a) 2 L
    (b) 250 mL
    (c) 600 mL

12. Repeat Question 11 for the most economical cylindrical containers.

13. Unfold a cereal box.
    (a) What do you notice?
    (b) Determine the actual amount of material used.
    (c) Calculate its economy rate based on the actual amount of material used.

14. Suppose plastic costs 0.015¢/cm². Atlantic Packaging has been asked to design a rectangular-prism container with the maximum capacity possible for a total cost of 5¢.

    Describe the container. Explain what you did to find the dimensions.

15. (a) Which do you think has the greater economy rate, a 250-mL milk carton or a 1-L milk carton? Explain your prediction.

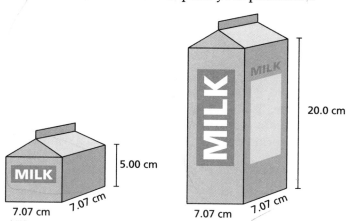

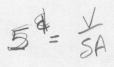

$5¢ = \dfrac{V}{SA}$

(b) Calculate the economy rates.
(c) What do you notice? Why do you think this is the case?

Project Connection

The traditional homes of many cultures are round. Round houses are warmer in cold climates and cooler in warm climates. They can also withstand extremely windy conditions.

The Inuit people of Canada traditionally used blocks of snow to build igloos in the shape of a hemisphere, or a half sphere. The thick snow walls and the round shape would help to hold heat inside the igloo.

1. Suppose an igloo is 4.6 m in diameter. Determine its surface area and volume.

2. Suppose you built a snow house with a square floor, flat walls, and flat roof. Find its surface area if it has the same height and volume as the igloo above.

3. What is the economy rate for the two houses in Question 2? Use this information and the fact that heat loss is proportional to surface area to explain why hemispherical houses would be best for the Arctic.

4. Determine the surface area of each container. The containers are designed to contain one sphere with a diameter of 8.0 cm. The cross section of each figure is a regular polygon.

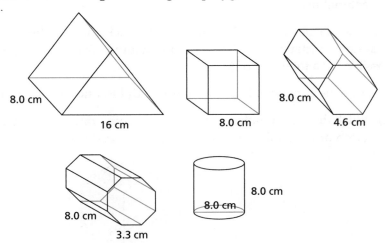

5. Using the volumes that you calculated for these containers in Section 6.2, determine the economy rate for each container.

6. Which container is the most economical? Explain.

6.5 Similarity and Size

Many products come in several different sizes. Larger sizes often cost less by volume or mass but they are harder to handle and may become stale before being completely used. Smaller sizes are more convenient, less likely to spoil or go stale, and easier to store for the individual consumer, but they are often more expensive. In this section, you will investigate the relationship between the size of a container and its economy rate.

Investigation 6
Making an Enlargement by Construction

Purpose
To construct an enlarged version of a picture and then compare side lengths, angle measures, and area.

Procedure

A. Draw a regular polygon. Choose any point O outside the polygon. Draw rays from O through each vertex of the polygon.

B. Use your ruler to find the length from O to one of the vertices, A. Double this number and use it to place the image of that vertex, A′, on the ray OA so that OA′ = 2(OA). Use the same process to find the image points for the other vertices of the polygon. Connect the image points to form the image polygon.

C. Use a protractor to compare the corresponding angles in your two polygons. What do you notice?

D. Use a ruler to compare the corresponding lengths in the two polygons. What do you notice?

E. Calculate the area of each of your polygons. What do you notice?

F. Choose another point, P, inside the polygon. Repeat Steps A to E using P instead of O. How is your image polygon different? How is it the same?

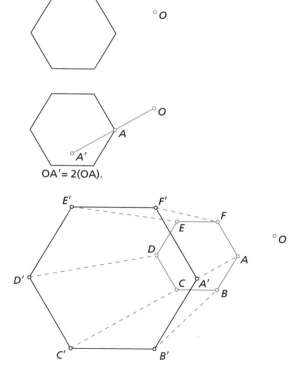

Check Your Understanding

1. European paper has dimensions in the ratio $1:\sqrt{2}$. For example, letter-size paper, called A4, is 210 mm by 297 mm. Each size can be enlarged to the next size by using a scale factor of $\sqrt{2}$.
   (a) A3 is the paper size one larger than A4. What are the dimensions of A3?
   (b) How could you determine the dimensions of A5 paper? *divide* A8 paper?
   (c) What are the dimensions of A1 paper? Explain.
   (d) How do the areas of A4 and A5 paper compare? How do the areas of any two consecutive paper sizes compare?

2. Suppose you wish to reduce an original drawing that is 10.0 cm wide and 8.2 cm high. It will have to fit in a 4.6-cm by 3.2-cm space in your school newspaper. What scale factor should you use on the photocopier? The copier has scales from 25% (scale factor of 0.25) to 120% (scale factor of 1.2).

3. Draw a diagonal across a rectangular sheet of paper. Mark any point on the diagonal. Make horizontal and vertical folds through that point. Unfold the paper.
   (a) The two folds divide the original piece of paper into four rectangles. Which of these rectangles are similar to the original piece of paper?
   (b) Determine the scale factor of reduction for each similar rectangle.
   (c) What proportion of the original rectangle's area is the area of each similar rectangle?
   (d) If you wished to create a rectangle that was $\frac{1}{5}$ the scale of the original, where would you place your point on the diagonal?

Similar Shapes and Scale Factor

How can you determine if a pair of two-dimensional shapes are similar? A pair of two-dimensional shapes are similar if they have the same shape and the lengths of their corresponding sides have the same ratio or scale factor.

Example

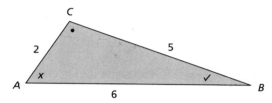

$$\frac{AB}{DE} = \frac{BC}{EF} = \frac{CA}{FD}; \frac{6}{3} = \frac{5}{2.5} = \frac{2}{1} = 2$$

The scale factor is 2.

How can you determine if two three-dimensional shapes are similar?

Two three-dimensional shapes are similar if they have the same shape and the lengths of the corresponding edges have the same ratio or scale factor.

Example 1

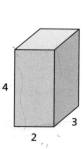
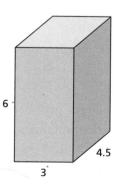

These shapes are similar because $\frac{2}{3} = \frac{4}{6} = \frac{3}{4.5}$. The scale factor is $\frac{2}{3}$.

6.5 Similarity and Size

Example 2

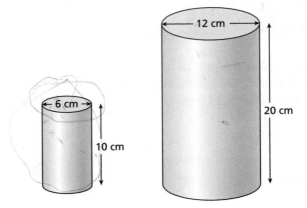

These shapes are similar because $\frac{6}{12} = \frac{10}{20} = 0.5$. The scale factor is 0.5.

How can you create similar three-dimensional shapes?

You can create similar shapes by multiplying all the dimensions of one shape by the same factor.

Example 3
To create a larger similar prism, all three dimensions are multiplied by the scale factor, 3.

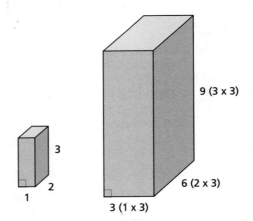

Similar Shapes
  (a) Create a larger prism similar to those shown above.
  (b) Now create a smaller similar prism.

Focus Question

4. Examine two sizes of the same product. Are the containers similar? Explain what you did to determine similarity.

Investigation 7
Creating Similar Shapes

Purpose

To investigate how increasing the size of a rectangular prism affects its volume, surface area, and economy rate.

> **Did You Know?**
>
> Ice cream was first served on a cone at the St. Louis World's Fair in 1904. When an ice cream vendor ran out of paper dishes, he rolled waffles into a cone shape and placed the ice cream on top.

Procedure

A. Use connecting cubes to investigate how increasing all dimensions of a rectangular prism by the same scale factor affects its volume, surface area, and economy rate.

Record your data in a table as follows.

n n^3 n^2

Original Dimensions (cm)	Scale Factor, n	New Dimensions (cm)	Original Volume, V_0 (cm³)	New Volume, V_1 (cm³)	Ratio, $\frac{V_1}{V_0}$	Original Surface Area, SA_0 (cm²)	New Surface Area, SA_1 (cm²)	Ratio, $\frac{SA_1}{SA_0}$	Economy Rate
1.00 × 1.00 × 1.00	2	2.00 × 2.00 × 2.00	1.00	8.00	8	6.00	24	4	.166 .333
1.00 × 1.00 × 1.00	3	3.00 × 3.00 × 3.00	1.00	27.0	27	6	54	9	.6 .5
1.00 × 1.00 × 1.00	4	4×4×4	1	64	64	11	96	16	.166 .6
2.00 × 2.00 × 2.00	2	4×4×4	8	64	8	24	96	4	.3 .6
2.00 × 2.00 × 2.00	3	6×6×6	11	216	27	11	216	9	11 1
2.00 × 2.00 × 2.00	4	8×8×8	11	512	64	11	384	16	11 .75
3.00 × 3.00 × 3.00	2	6×6×6	27	216	8	54	216	4	.5 1
4.00 × 4.00 × 4.00	½	2×2×2	64	8	0.125	96	24	.25 .66 .333
4.00 × 4.00 × 4.00	¼	1×1×1	64	1	.1563	96	6	.625 .66 .166

B. What do you notice about the relationship between the scale factor, n, and the ratio of the volumes? Surface areas?

C. Use your data to investigate how the scale factor and the economy rate of similar containers are related.

D. What effect does a scale factor of less than 1 have on:
(a) volume?   (b) surface area?   (c) economy rate?

E. Summarize, in writing, what you have learned about the relationship between scale factor, n, and:
(a) volume;
(b) surface area; and
(c) economy rate.

Handwritten notes:
- Scale Factor is $\sqrt[3]{\text{Volume Ratio}}$
- Scale Factor is $\sqrt{\text{SA Ratio}}$
- $\frac{\text{Scale Factor}}{ER}$

Investigation Questions

5. Suppose that all the dimensions of this soup can were multiplied by a scale factor of 1.5. Predict its volume using what you know about the relationship between scale factor and the ratio of volumes. Calculate its volume to check.

capacity: 283 mL
SA: 245 cm²

6. Predict the surface area of the 283-mL soup can if all dimensions are multiplied by the scale factor 1.5. Calculate to check.

7. Copy and complete the table.

Container	V_1	SA_1	ER_1	Scale Factor	V_2	SA_2	ER_2
(a) Cylinders $d = 8.00$ cm $h = 12.0$ cm	603.2	402	1.5	2	4824	608	3
(b) Rectangular Prisms $w = 10.0$ cm $d = 5.00$ cm $h = 14.0$ cm	700	520	1.35	0.5	87.5	130	.67

8. Compare the economy rates for each shape in Question 7 by finding $\dfrac{ER_2}{ER_1}$. This is called the ratio of the economy rates. Does this comparison have any significance for packaging design?

Check Your Understanding

9. Typically, different-sized containers for the same product are not similar although scale factors are still involved. Explain what this means using the example shown.

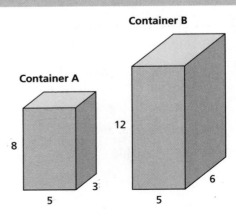

10. (a) Find the volume of Container A in Question 9.
    (b) Predict the volume of Container B using Container A's volume and what you know about the scale factors.
    (c) Calculate Container B's volume. How does it compare to your prediction?
    (d) Summarize, in writing, the relationship between the volumes of two shapes if their dimensions differ by different scale factors.

★ 11. Ken argues that Container C has a greater economy rate because it's a cube. Nicolas argues that Container D has a greater economy rate because its volume is greater. Who do you think is right? Why?

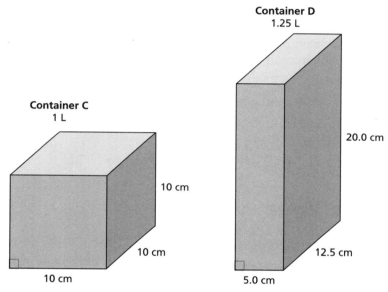

12. Emma claims that the container that is more cube-like always has the greater economy rate. Breanne claims that the container with the greater volume always has the greater economy rate. Both students are partly right.
    (a) Describe a situation where Emma would be right.
    (b) Describe a situation where Breanne would be right.

FOCUS G — Solving Problems Involving Scale Factors

If all the dimensions of Container A are multiplied by the same scale factor, n:
- the resulting Container B will be similar;
- the volume of Container B will be n^3 times the volume of Container A;
- the surface area of Container B will be n^2 times the surface area of Container A; and
- the economy rate of Container B will be n times the economy rate of Container A.

6.5 Similarity and Size   289

If the dimensions of Container A are multiplied by different scale factors, m, n, and p, the resulting Container B will have a volume that is mnp times the volume of Container A.

You can use these relationships to solve problems.

Example 4

Company A wants to package a 300.0-g box of cereal in a similar 600.0-g size. By what factor will they have to increase the dimensions to get a larger similar box?

Knowing that the package is 300.0 g tells us how much the package weighs, not the volume of the container's contents. If the weight of a product doubles, its volume will also double since weight and volume are directly related. Thus, if the company wishes to package its product in a 600.0-g box instead of a 300.0-g box, the volume of the package will also double. Thus $V_2 = 2V_1$.

The boxes must be similar so we know: $V_1 \times n^3 = V_2$

Substitute $2V_1$ for V_2: $V_1 \times n^3 = 2V_1$

Solve for n: $n = \sqrt[3]{\dfrac{2V_1}{V_1}}$

Think about...

Product Packages

Why do you think different sizes of the same product are not usually packaged in similar containers?

Check Your Understanding

13. Company B wants to package their 500.0-g box of cookies in a similar 150.0-g size. By what factor will the dimensions change?

14. All dimensions of a 125-mL cubic carton are tripled. What is the:
    (a) resulting volume?   (b) resulting surface area?

15. Determine the volume of the box that would result from doubling one dimension and tripling the other two dimensions.

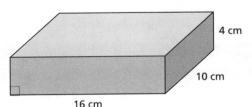

16. Scaling a shape results in a greater volume. Do you agree or disagree? Explain your reasoning.

17. Suppose you want to increase the volume of a box by a factor of 12. List three different ways you could do it.

18. A cereal manufacturer has approached Atlantic Packaging about designing two boxes for a new line of cereal in sizes 400 g and 1000 g. They want the design of the new product to attract attention. Write a proposal that outlines what you would suggest. Provide diagrams and your rationale. Note: Each 200 g of cereal has a volume of 1000 cm^3.

Project Connection

1. Sphere A has a radius of 12 cm; sphere B a radius of 8 cm.
   (a) What is the ratio of their diameters?
   (b) What is the ratio of their surface areas?
   (c) What is the ratio of their volumes?

2. Explain what effect doubling the diameter of a sphere has on its surface area and its volume.

3. The radii of five spheres are 1.0 cm, 2.0 cm, 3.0 cm, 4.0 cm, and 5.0 cm.
   (a) Find the volume and the surface area for each sphere.
   (b) Determine the economy rate for each sphere.
   (c) Graph the economy rate as a function of radius. What do you notice?

4. Weather balloons come in various sizes. Larger balloons can lift more equipment than smaller balloons. The table below shows the mass that can be lifted by two different balloons.

Diameter of Balloon	Mass Balloon Can Lift
2.4 m	7.7 kg
4.8 m	62.1 kg

   (a) Assume the balloons are spheres. Find the volume of each.
   (b) Compare the ratio of the diameters to that of the volumes.
   (c) For each balloon, find the rate comparing the volume and the mass that the balloon can lift. What do you notice?
   (d) How can you estimate the mass that a 8-m diameter balloon can lift?

Think about...

Giants
Why is there a limit to the size of the human body? Why can't giants exist? How can you use scaling in your explanation?

6.6 Variations in Packaging

In the first five sections of this chapter, the focus was on designing the most economical container possible. This section considers other factors that influence packaging.

Focus H: Some Packaging Considerations

Think about...

Rigidity

A triangular-prism-shaped poster tube is more rigid than a cylindrical tube made of the same material. Why do you think this is the case?

If triangular tubes are stronger, why aren't they used more often to ship posters?

inscribed circle – a circle that touches each side of the triangle

incentre – the centre of the inscribed circle

circumscribed circle – a circle that passes through each vertex of the triangle

circumcentre – the centre of the circumscribed circle

The rigidity of a shipping container is a factor in keeping shipping costs down. If products are shipped in containers that are not rigid enough, they have to be handled more carefully, thus increasing handling costs. Also, damage is more likely, thus increasing costs even more.

For example, rigidity considerations might prompt a manufacturer to package rolled posters in prism-shaped containers with triangular cross-sections. Such a situation might be described mathematically as a circle **inscribed** in a triangle.

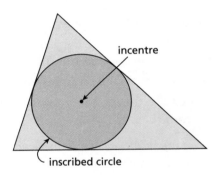

In a similar way, a company such as Curious Candles might decide to package and market its triangular prism-shaped candles in clear plastic, cylindrical containers, as shown in cross-section below left.

As shown at the far right, a circle is **circumscribed** about a triangle if every vertex or corner of the triangle lies on the circle. The centre of the circumscribed circle is the **circumcentre** of the triangle.

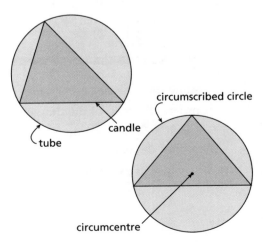

Focus Question

1. (a) Draw a large, acute, scalene triangle.
   (b) Using a compass, draw its circumscribed and inscribed circles. Explain what you did.

Investigation 8
Exploring Triangles

Purpose

To explore some of the characteristics of a triangle that relate to the packaging situations discussed in Focus H.

Procedure

A. Draw four large, congruent, acute, scalene triangles.

B. Use one of the triangles and draw the **perpendicular bisectors** of all three sides. What do you notice?

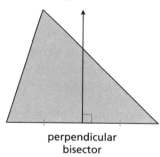

perpendicular bisector

C. Using another triangle, draw the **medians** of each side. What do you notice?

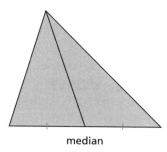

median

D. Using a third triangle, find the three **angle bisectors**. What do you notice?

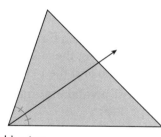

angle bisector

> **Did You Know?**
> Computer software is available to carry out investigations like this one interactively.

perpendicular bisector – a line that is perpendicular to and passes through the midpoint of a side of a triangle

median – a segment from a vertex to the midpoint of the opposite side of a triangle

angle bisector – a segment that bisects or cuts an angle into two congruent parts

altitude – a perpendicular segment from a vertex of a triangle to the line that contains the opposite side

Think about...

Step F
How could you use what you have learned to find the centre of a circle?

coincident – occurring in the same place

Did You Know?

Doctors use CAT scans to get cross-sectional images of the body. CAT stands for Computerized Axial Tomography. Tomography comes from *tomos*, the Greek word for section or cut.

E. Using your fourth triangle, draw the three altitudes. What do you notice?

F. Use your constructions in B to E. Is the intersection point of the perpendicular bisectors, the medians, the angle bisectors, or the altitudes the centre of the inscribed circle of a triangle? Explain what you did to find out.

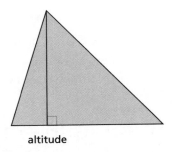
altitude

G. Which of the intersection points in Question F is the centre of the circumscribed circle of a triangle? Explain what you did.

H. Which of the intersection points in Question F do you think is the centre of gravity of a triangle? What could you do to check?

I. Repeat the Investigation using an equilateral triangle. What do you notice?

Investigation Questions

2. Are the points determined in Steps B, C, D, and E all the same, some the same and some different, or all different? Be prepared to clearly justify your answer.

3. Consider the points determined in Steps B and D. Write a valid argument to prove that the points are not coincident.

4. In which situation are the intersection points of the perpendicular bisectors and angle bisectors (as in Steps B and D, respectively) coincident? Are these the only intersection points that are coincident in this situation?

Check Your Understanding

5. (a) Find the circumscribed and inscribed circles and centre of gravity for an acute isosceles triangle, an obtuse isosceles triangle, and a right isosceles triangle.
   (b) Examine each figure. What conclusion might you reach concerning the angle bisector of the vertex angle of each isosceles triangle and the perpendicular bisector of the opposite side?
   (c) Write a valid argument to prove the conclusion you reached in part (b).

6. Draw a non-equilateral triangle that represents the open end of a prism-shaped shipping tube. To what diameter would a poster have to be rolled to fit inside? Explain what you did.

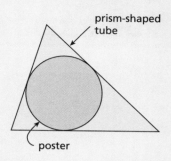
prism-shaped tube

poster

7. How would you determine the inscribed and circumscribed circles for any regular polygon?

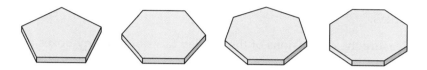

8. Every polygon has a centre of gravity. It is the point at which the figure "balances." For regular polygons, such as squares and equilateral triangles, this point is easy to estimate visually. For non-regular polygons, the point is less obvious. If you were to cut out a cardboard triangle, you could locate its centre of gravity by finding the point at which the triangle balances on a pinhead. Is it important to consider the location of the centre of gravity when designing a package? Why or why not?

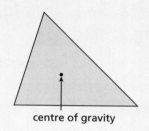
centre of gravity

Project Connection

If spheres are to be packaged in a container, they will sometimes be packaged in a single row or column, as in cans of tennis balls. At other times they will be piled together to take advantage of their nesting qualities.

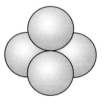

1. Spheres are to be packaged in a square-based prism. Sketch a diagram that illustrates the way you think they could be most effectively packed.

2. Spheres are to be packaged in a cylinder. Sketch a diagram that illustrates the way you think they could be most effectively packed.

3. The tetrahedron is one of the strongest shapes for a package. Sketch a diagram that illustrates the way you think spheres could be packaged to take advantage of this container shape.

— Note —
Now you are ready to finish your chapter project. Your completed project should include:
- detailed and well-organized solutions to all the questions labeled project connection.
- the container you designed to hold tennis balls.
- a set of drawings that represent the 3-D shape of the container with enough detail for another person to construct it.
- a clear explanation defending your choice of material, shape, and dimensions.

PUTTING IT TOGETHER

CASE STUDY 1

Find patterns in the ratios of the dimensions of the display faces of containers.

1. Measure the dimensions of the display faces of as many boxes as you can.

2. Graph the relationship between the longer and shorter dimensions of each face. Place the shorter dimension along the horizontal axis.

3. Plot points until you can determine a regression line for the data. What is the slope?

4. Repeat for the height and diameter of cylindrical containers.

5. Research the golden ratio. Does it apply to the ratios you see in the dimensions of display faces? Explain.

6. Write a report about what you discovered.

CASE STUDY 2

For a company that uses millions of disposable fast-food containers in a year, every penny saved by economy of container design is significant. A fast-food chain sells hundreds of thousands of beverages daily. The beverage containers are designed with a great deal of thought.

1. Compare the economy rates of various-sized beverage cups from a fast-food restaurant.

2. Determine the rate for the cost of the beverage to the capacity for each size of cup. Compare the rates for the different sizes of cups.

3. Write a report summarizing what you discovered.

4. Redesign each size of cup so that it holds the same amount using less material. Present your new designs using diagrams. Highlight the advantages and disadvantages of each.

5. Design a container that will hold two dozen cups of one size (small, medium, or large). Include a diagram of your container and explain your rationale for the design.

REVIEW

Key Terms

	page
altitude	294
angle bisector	293
apothem	267
capacity	262
centre of gravity	295
circumcentre	292
circumscribed circle	292
coincident	294
cube root	278
economy rate	275
exterior angle	265
hendecagon	269
heptagon	264
incentre	292
index	279
inscribed circle	292
interior angle	265
lateral face	270
lateral surface	270
line symmetry	265
median	293
nonagon	267
order of rotational symmetry	265
pentagon	264
perfect cube	279
perfect square	278
perpendicular bisector	293
pyramid height	259
radicand	279
regular polygon	264
rigidity	292
rotational symmetry	265
scale factor	285
slant height	270

You Will Be Expected To

- use previously learned geometric and measurement concepts and formulas to solve problems.
- calculate the area of any regular polygon with a known perimeter and number of sides.
- calculate the volume and surface area of prisms, cylinders, pyramids, cones, and spheres.
- calculate the economy rate of any shape by dividing its volume by its surface area.
- compare the economy rates of different shapes to determine the most economical design for a given volume.
- determine if two three-dimensional shapes are similar by comparing the ratios of their corresponding edges.
- determine the scale factor between two similar shapes by comparing their volumes or their surface areas.
- use the scale factor to determine the volume and surface area of a similar shape.
- locate the medians of any triangle and any regular polygon and use them to find the centre of gravity.
- locate the perpendicular bisectors any triangle and of any regular polygon and use them to locate the circumscribed circle.
- locate the angle bisectors of any triangle or any regular polygon and use them to locate the inscribed circle.
- locate the altitudes of any triangle.

Summary of Key Concepts

6.1 Examining Factors in Container Design

The volume of any prism or cylinder can be calculated using the formula $V = Ah$ (where A is the area of the cross-section or base and h is the height of the prism or cylinder). This is because the area of the cross-section is the same over the entire height.

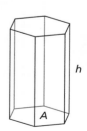

The volume of a pyramid is one third the volume of a prism with the same base and height.

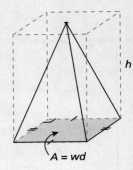

The volume of a cone is one third the volume of a cylinder with the same base and height.

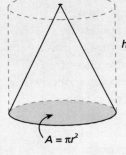

The volume of a sphere is given by $\frac{4}{3}\pi r^3$, where r is the radius of the sphere.

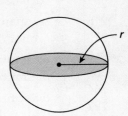

6.2 Regular Polygons

The area of a regular polygon can be calculated if you know its perimeter or side length and the number of sides.

Let s represent the length of one side of the polygon, and n represent the number of sides of the polygon. The interior angle at the vertex of the polygon is $\frac{180°(n-2)}{n}$.

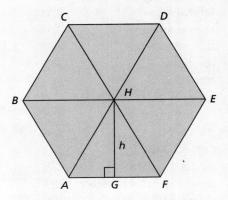

$$\tan \angle HAF = \frac{h}{AG}$$

$$h = AG \times \tan \angle HAF$$

$$= \frac{s}{2} \times \tan \frac{180°(n-2)}{2n}$$

Area of $\triangle HAF$

$$= \frac{1}{2} AF \times h$$

$$= \frac{1}{2} s \times \frac{s}{2} \times \tan \frac{180°(n-2)}{2n}$$

$$= \frac{s^2}{4} \tan \frac{90°(n-2)}{n}$$

Area of polygon

$$= n \times \text{Area of } \triangle HAF$$

$$= \frac{ns^2}{4} \tan \frac{90°(n-2)}{n}$$

298 Chapter 6 *The Geometry of Packaging*

Example
This pentagon has side lengths of 20.0 cm. Find the area.

Solution
Area of the pentagon

$= \dfrac{ns^2}{4} \tan \dfrac{90°(n-2)}{n}$

$= \dfrac{(5)(20.0)^2}{4} \tan \dfrac{90°(5-2)}{5}$

$= 500 \tan 54°$

≈ 688

The area is 688 cm².

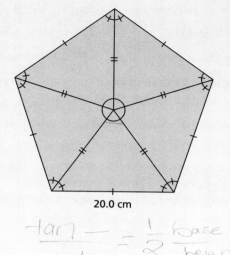

20.0 cm

For a given perimeter, as the number of sides of a regular polygon increases, the area increases.

P = 28 cm, A = 49 cm²   P = 28 cm, A = 56 cm²

For a given area, as the number of sides of a regular polygon increases, the perimeter decreases.

A = 49 cm², P = 28 cm   A = 49 cm², P = 26 cm

6.3 Surface Area

The surface area of a three-dimensional object with polygonal faces is the sum of the areas of the faces.

The surface area of a cone is given by $\pi r^2 + \pi rs$, where r is the radius of the base and s is the slant height.

The surface area of a sphere is given by $4\pi r^2$, where r is the radius of the sphere.

The surface area of a cylinder is given by $2\pi r^2 + 2\pi rh$, where r is the radius of the base and h is the height of the cylinder.

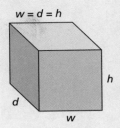

6.4 Economy of Design

The economy rate of a shape can be calculated by dividing its volume by its surface area: $ER = \dfrac{V}{SA}$. When comparing containers with the same volume, a greater economy rate means there is less surface area, or less material is required.

For a given volume, the rectangular prism with the greatest economy rate is a cube.

For a given height and volume, as the number of sides of a regular polygon-based prism increases, the surface area decreases and therefore the economy rate increases.

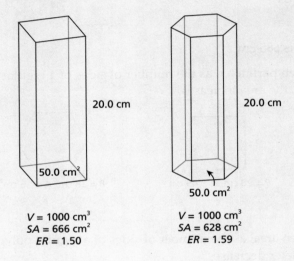

$V = 1000 \text{ cm}^3$
$SA = 666 \text{ cm}^2$
$ER = 1.50$

$V = 1000 \text{ cm}^3$
$SA = 628 \text{ cm}^2$
$ER = 1.59$

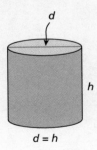

For a given volume, the cylinder with the greatest economy rate has an equal diameter and height.

For a given volume, a cylinder with equal diameter and height is more economical than a cube.

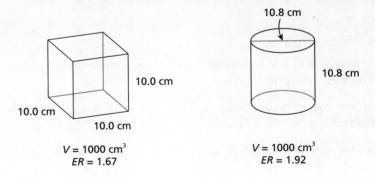

$V = 1000 \text{ cm}^3$
$ER = 1.67$

$V = 1000 \text{ cm}^3$
$ER = 1.92$

6.5 Similarity and Size

When all dimensions of a prism or cylinder are multiplied by the scale factor n:

- the resulting shape is similar;
- its volume changes by a factor of n^3;
- its surface area changes by a factor of n^2; and
- its economy rate changes by a factor of n.

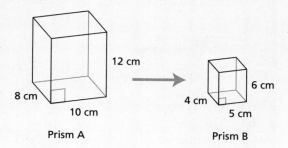

Prism A   Prism B

Scale factor to change Prism A to Prism B is 0.5.
$\left(\dfrac{5}{10} = \dfrac{4}{8} = \dfrac{6}{12}\right)$.

	Prism A	Prism B
Volume	960 cm³	$(0.5)^3 \times 960 = 120$ cm³
Surface Area	592 cm²	$(0.5)^2 \times 592 = 148$ cm²
Economy Rate	1.62	$0.5 \times 1.62 = 0.81$

If two shapes are similar, the one with the greater volume has the greater economy rate.

When the dimensions of a prism or cylinder are multiplied by different scale factors, m, n, and p, the volume changes by a factor of the product of the scale factors, mnp.

6.6 Variations in Packaging

The three perpendicular bisectors of any triangle or regular polygon intersect at one point. This intersection point is the centre of the circumscribed circle or the circumcentre of the triangle or polygon.

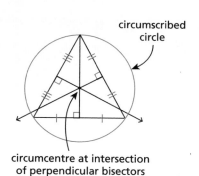

circumscribed circle

circumcentre at intersection of perpendicular bisectors

The three angle bisectors of any triangle or regular polygon intersect at one point. This intersection point is the centre of the inscribed circle or the incentre of the triangle or polygon.

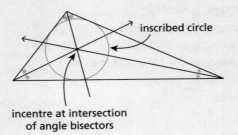

incentre at intersection of angle bisectors

The three medians of any triangle intersect at one point. This intersection point is the centre of gravity.

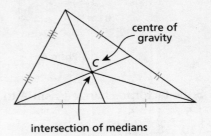

intersection of medians

The three altitudes of any triangle intersect at one point.

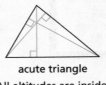 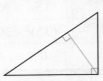 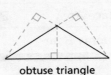

acute triangle
All altitudes are inside.

right triangle
Two legs are altitudes.

obtuse triangle
Two altitudes are outside.

PRACTICE

6.1 Examining Factors in Container Design

1. Find the capacity of this Blue Box, which consists of the shaded part of the square-based pyramid.

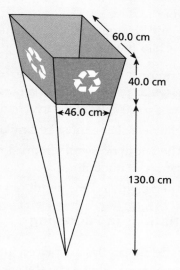

2. Explain how you would determine the volume of this container. Explain why your method works.

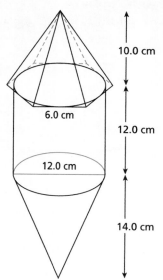

3. (a) Find 3 possible sets of dimensions for a 1-L juice box.

   (b) Find an actual 1-L juice box. Do any of your sets of dimensions match the actual dimensions?

6.2 Regular Polygons

4. Determine the unknown angles in each regular polygon.
   (a)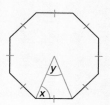
   (b)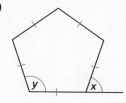

5. Find the area of each regular polygon.
   (a)
   (b)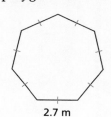

6. (a) A regular hexagon and a regular octagon both have the same perimeter. How do their areas compare? Explain your answer.

   (b) A square and an equilateral triangle both have the same area. How do their perimeters compare? Explain your answer.

6.3 Surface Area

7. Determine the surface area of each.
   (a)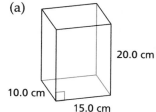
   (b)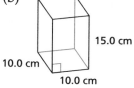

7. continued

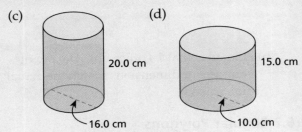

(c) 20.0 cm, 16.0 cm (d) 15.0 cm, 10.0 cm

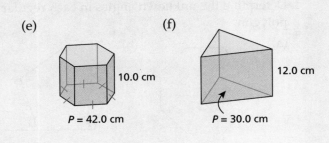

(e) 10.0 cm, P = 42.0 cm (f) 12.0 cm, P = 30.0 cm

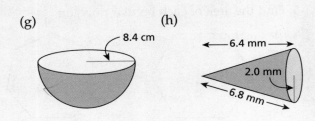

(g) 8.4 cm (h) 6.4 mm, 2.0 mm, 6.8 mm

6.4 Economy of Design

8. Find the economy rate for each container in Question 7.

9. A container holds 1500 mL.
   (a) Describe the most economical rectangular prism container possible with a height of 15 cm. Defend your choice.
   (b) Describe the most economical rectangular prism container. Defend your choice.
   (c) Is there another kind of prism container that is more economical? Explain.
   (d) What are the dimensions of the most economical container?

6.5 Similarity and Size

10. This cylinder has a capacity of 1810 mL, a surface area of 829 cm², and an economy rate of 2.18. Determine the volume, surface area, and economy rate for a similar cylinder that is half the height.

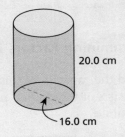

20.0 cm, 16.0 cm

11. Determine how the dimensions of a 500-g rectangular prism would change to decrease its mass to 300 g under each condition:
    (a) The new shape must be similar.
    (b) The width or the depth must remain the same but not both.
    (c) The height and the depth cannot change.

6.6 Variations in Packaging

12. Cylindrical candles are placed into triangular prism-shaped tubes to be shipped.
    (a) Determine the maximum diameter of each candle that would fit inside each tube. For each, explain what you did.
    (b) Which tube would you recommend? Why?

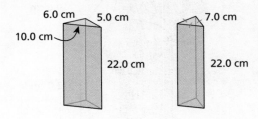

6.0 cm, 10.0 cm, 5.0 cm, 22.0 cm; 7.0 cm, 22.0 cm

13. (a) Draw an equilateral triangle.
    (b) Draw its circumscribed circle.
    (c) What percent of the total area of the circle is taken up by the triangle?
    (d) Repeat for several other equilateral triangles. What do you notice?

Chapter Seven
Linear Programming

Linear programming was developed after World War II by economists, mathematicians, and others to help industries choose the best solutions to manufacturing problems. This was necessary because of the lack of money and industry (in war-torn Europe) at the end of World War II.

One of the problems studied was which cost-cutting measures would save the manufacturing industry the most money under different constraints.

For example, the oil industry wanted to know the most efficient way of refining and blending crude oil to produce the highest quality oil at the least expensive price. In today's world, it has been estimated that up to 10% of all computer time in the natural resource industry is spent solving linear programming-like problems such as this.

After successfully completing this chapter, you will be expected to:

1. Identify possible and reasonable constraints for variables in problem situations and represent these algebraically and graphically.

2. Solve systems of equations using:
   • a graphical method.
   • an algebraic method.

3. Use the process of linear programming to find the optimal solution to a problem represented by linear relationships in two variables.

7.1 Exploring an Optimization Problem

— Note —
To help you solve a problem like this, you could imagine yourself in Heather's place and answer the questions. This is a useful problem-solving technique.

Heather was looking for a part-time job so she could save money to buy a car. She found a job cutting material for Spinney Manufacturing, a company that makes upholstered furniture. Her job is to cut material for couches and chairs. Heather is paid a flat fee for each bundle of fabric she cuts. A bundle includes all of the different pieces of fabric that are sewn together to make a chair or couch.

Heather is paid $13.50 for every couch bundle she cuts and $5.50 for every chair bundle. If she completes only a part of a bundle, she earns only that fraction of the money. For example, since she is paid $5.50 for each chair bundle, she is paid $2.75 if she cuts half of a chair bundle. It takes her 2 h to cut fabric for an entire couch and 45 min to cut fabric for an entire chair.

What is the greatest amount of money Heather can earn?

Investigation 1
Factors That Affect Heather's Income

Purpose
Investigate the variables in a problem and the relationships between the variables.

$27.50 — 5 chairs
$27.00 — 2 couches

Procedure

13.50n + 5.50n = income

A. Try different combinations of couch and chair bundles that can be cut. What is the maximum number of chair bundles Heather should cut to earn the maximum amount of money? What is the maximum number of couch bundles she should cut? List all of the assumptions you made.

B. How can Heather calculate the quantity she is trying to maximize that you identified in Step A? What equation can be used to describe this relationship? If you cannot develop an equation, explain why.   *13.50x + 5.50y = income*

constraint – a restriction on the allowable values of a variable in a problem

C. List at least five **constraints** that you think are important in solving the problem. Give reasons for each. Calculate the income Heather can expect when all five of the new constraints are considered.

306   Chapter 7 *Linear Programming*

Investigation Questions

1. Based on your values when you use the constraints for the chair bundles and couch bundles cut, what is the maximum income Heather can earn?

2. What number of couch and chair bundles would be cut by Heather using your constraints? If you were Heather's manager at Spinney Manufacturing, would you be pleased with the number of couch bundles and chair bundles she plans to cut? Why?

3. Is the number of chair bundles cut dependent on the number of couch bundles cut? Justify your answer.

Heather's Problem
Is there a single correct answer to Heather's problem? Why or why not?

Focus A — Heather's Personal Constraints

Heather decided that in every 24-h day during the week, she would spend her time as shown below:

- 7 h sleeping;
- 1 h eating;
- 2 h relaxing, watching TV, or being with her friends;
- 1 h exercising, hiking, riding a bike, or other physical activities;
- 6.5 h at school;
- 1.5 h completing her homework; and
- the rest of the time would be spent working for Spinney Manufacturing. 5hrs

—Note—
The list shows the constraints Heather puts on her day.

Focus Questions

4. Refer to Investigation 1. Repeat Step A using Heather's list of constraints. How are the results like the ones you got previously? How are they different?

5. What is the maximum amount of money Heather can earn in one week under her constraints? How do you know?

6. A list of constraints Heather imposes on herself is given in this Focus. Refer to your work in the Focus and the list.

   (a) Suppose Heather wanted to sleep 8 h per night instead of 7 h per night. How might this affect how much she can earn?

   (b) Suppose Heather decided she will get enough exercise working and will not need the 1 h per day of exercise she planned. How might this affect how much she can earn?

Think about...

Heather's Constraints

Are there any other constraints Heather should consider? How might these affect the amount of money Heather could earn?

— Note —
Spinney Manufacturing must also consider the needs of its customers. They want to purchase both couches and chairs.

1500 lettuce
500 corn

1000 lettuce
1000 corn

1200 lettuce
1300 corn

Think about...

Step A

Compare constraints you listed with those listed by others in the class. Did others list constraints that you did not consider? Are there other constraints that should be considered?

(c) Suppose you were considering a job at Spinney Manufacturing.

(i) Make a list of the constraints you would have for yourself.

(ii) Based on your list, what maximum income could you earn?

(iii) Are there any constraints you listed for yourself that you think Heather should also consider? What are they? Why should Heather list them?

Investigation 2
The Manufacturer's Constraints

Purpose

To consider the constraints that the manufacturer needs to place on Heather's part-time job.

Procedure

A. Brainstorm a list of constraints that Spinney Manufacturing might place on Heather's work day. Give reasons for each constraint.

B. Decide which of the constraints from Step A should be placed on Heather's work day. For these constraints, set some realistic numerical values.

- What is the maximum number of chair bundles Heather should cut to earn the maximum amount of money?
- What is the maximum number of couch bundles she should cut?

Check Your Understanding

7. Round the Bend Farm is a 2000-ha vegetable farm. Sue, the owner, can plant no more than 1500 ha of lettuce and no more than 1000 ha of corn to satisfy customer demand.

(a) List Sue's constraints.

(b) How many hectares of each crop can she plant? Compare your answers with those of others in the class. Are the answers the same? Why is it likely that they are not the same?

(c) Write an answer for this problem that is not possible based on the constraints given. Explain why the answer is impossible.

(d) What piece of additional information about the farm might help you get the same answer as others in the class for part (b)? Why would this piece of information help?

8. The Sunny Garden Gift Shop is going to package holiday gift packs of jam and marmalade. They have 50 jars of marmalade and 90 jars of jam. The Morning Glory package will contain four jars of marmalade and one jar of jam. The Berry Patch contains four jars of jam and one jar of marmalade. The company will earn $3 profit on the Morning Glory package and $4 profit on the Berry Patch.

   (a) If only Berry Patch gift packs were made, how many could be produced? How much profit would the company earn?

   (b) If only Morning Glory packages were made, how many could be produced? How much profit would the company earn?

   (c) Is there a combination of Morning Glory and Berry Patch packages that can be made that will earn the company more than the amount in (a) or (b)? If so, what is it? Is all of the available jam and marmalade used in your combination?

9. For each of the following problems:

   (a) Decide whether you have enough information to solve it.

   (b) List the constraints for the problem.

   (c) If the problem can be solved, solve it. Decide whether you have a unique solution. If the problem can't be solved, explain why.

   A. Manufacturing

   A firm manufactures bicycles and tricycles. The profit on a bicycle is $50 and the profit on a tricycle is $30. Find the maximum profit the manufacturer can make.

   B. Sporting Goods Manufacturing

   A sports-equipment manufacturer makes basketballs and soccer balls. Two different machines are used to make the balls and the time required by each machine is shown.

	Time on Machine A	Time on Machine B
Basketball	2 min	1 min
Soccer Ball	1 min	2 min

   The profit on a basketball is $1.95 and the profit on a soccer ball is $2.30. What is the maximum profit the company can make?

   C. Hockey Stick Manufacturing

   Two employees of a hockey stick manufacturer both make hockey sticks and goalie sticks. Janet can make 3 hockey sticks and 3 goalie sticks per hour. Sam can make 4 goalie sticks and 1 hockey stick per hour. The company makes $2.00 profit on each goalie stick and $1.00 profit on each hockey stick. How long should each person spend making hockey sticks and goalie sticks each day to minimize costs?

Think about...

Question 8(c)

Compare your answers with those of others in the class. What do you notice?

— Note —

The process you used to find the maximum profit can also be used to find the minimum cost. You will be considering expenses instead of a profit. Always list any assumptions you have made.

— Note —

A hockey stick manufacturer might normally sell more hockey sticks than goalie sticks. However, this trend can be reversed during the holiday seasons of any year.

7.1 Exploring an Optimization Problem 309

7.2 Exploring Possible Solutions

In her work at Spinney Manufacturing, Heather will choose to cut the combinations of chairs and couches for which she could make the most money. However, the company has to produce a certain number of couches and chairs each week to meet customer demand.

FOCUS B — Heather's Income

The manager of Spinney Manufacturing gave Heather the following constraints. She must:

- cut at least 10 couch bundles every two weeks;
- cut at least 8 chair bundles every two weeks;
- work a maximum of 18 h per week; and
- cut a maximum of 110 m of fabric every two weeks or the company will run out of fabric.

The design specifications for each piece of furniture indicate that:

- a chair requires 3 m of fabric to be cut; and
- a couch requires 5 m of fabric to be cut.

Think about...

Question 2
Why do you need to find Heather's income every two weeks and not every week or every month?

Focus Questions

1. Combine the above constraints with the ones listed by Heather in Focus A. Write a complete list of constraints for Heather's work.

2. Copy and complete the table to find Heather's income every two weeks. Complete the first two columns, using possible values for number of chair bundles cut and number of couch bundles cut, then calculate the income that Heather will earn from each combination. Find at least ten possible solutions that satisfy the constraints and at least ten that do not.

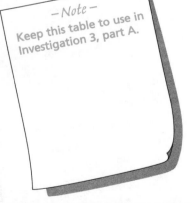

— Note —
Keep this table to use in Investigation 3, part A.

Couch Bundles	Chair Bundles	Income	Material Used	Time Needed	Constraints Satisfied?
10	8				
11.5	10				

310   Chapter 7 *Linear Programming*

3. How can you use the constraints to select possible solution values and eliminate others? Write a detailed description of how you determined whether certain values satisfied the constraints.

4. Are you closer now to finding Heather's maximum income every two weeks than you were before? Explain.

Investigation 3, Part A
Regions and Graphs in Heather's Problem

Purpose
To investigate how to graph an area in which all *feasible solutions* can be found.

Procedure

A. Refer to the table you created in Focus B. Get data from at least two other people to add to your table. You must have at least 30 pieces of data that satisfy the constraints and 30 that do not.

B. Plot all of your data from Step A on a coordinate grid where x is the number of chairs cut and y is the number of couches cut. Use red for the data that satisfy the constraints and blue for the data that do not satisfy the constraints.

C. Describe any pattern you see in the feasible data points and the non-feasible data points.

D. Is there an infinite number of feasible data points? Give reasons to support your answer.

Investigation Questions

5. Refer to Focus B on the previous page. Describe how the points on your coordinate grid are related to the constraints of Heather's problem.

6. Do you have all possible feasible solutions marked on your graph? Explain.

Think about...

The Chart
The process used in Question 2 to find possible solutions to Heather's problem is "guess and check." What are the limitations of using "guess and check" when solving this type of problem?

feasible solution – any solution to a problem that is possible within the constraints given

—Note—
The variable x is used to represent the number of chair bundles and y to represent the number of couch bundles. These will be used throughout the chapter.

—Note—
Plotting the feasible data points as a scatter plot will help you see any pattern.

Investigation 3, Part B
Regions and Graphs in Heather's Problem

— Note —
Remember that x represents the number of chair bundles cut and y represents the number of couch bundles cut as done in Investigation 3, Part A.

inequality – a mathematical statement that shows that two numerical or variable expressions are not always equal. For example, $3x \leq 12$ is an inequality.

— Note —
Spinney Manufacturing's constraints are not the only constraints in the problem. There are also the constraints on the amount of fabric that can be bought and how much time Heather has to work.

Purpose
To graph the points that represent the constraints in Heather's problem.

Procedure

A. Heather knew she had to cut at least 8 chair bundles every two weeks.
  - Write an **inequality** for this constraint.
  - Find at least 10 points that satisfy your inequality. Graph your points on a coordinate grid.
  - Add the points of three other students to your data. Plot these points on your coordinate grid.
  - Describe how the points show a region. Is there a "boundary" line for the region? If so, describe how you could identify the "boundary" line.

B. Heather also knew she had to cut at least 10 couch bundles every two weeks.
  - Write an inequality for this constraint.
  - Find at least 10 points that satisfy your inequality. Graph your points on a different coordinate grid.
  - Add the points of three other students to your data. Plot these points on your coordinate grid.
  - Describe how the points show a region. Is there a "boundary" line for the region? If so, describe how you can identify it.

C. Place your two coordinate grids on top of each other.
  - What does the overlap of the two regions represent?
  - Create a new graph with this overlap region on it.

Investigation Questions

7. Heather could see that the graph represented the number of chair bundles and couch bundles Spinney Manufacturing wanted her to cut. However, there were also constraints on how much fabric she could use.

    - 110 m of fabric in total is available every two weeks.
    - Each chair requires 3 m of fabric.
    - Each couch requires 5 m of fabric.

    (a) Write an inequality to connect these three constraints.

    (b) Find 10 points that satisfy the inequality. Show the points on a coordinate grid.

    (c) Place the grid from part (b) on top of the new grid you created in Step C. What does the overlap of these two regions represent? What information does this give Heather?

8. Heather then wanted to add the following constraints to the coordinate grid.

    - She can only work a maximum of 36 h every two weeks.
    - It will take 45 min to cut material for a chair.
    - It will take 2 h to cut material for a couch.

    (a) Write an inequality to connect these three constraints.

    (b) Find 10 points that satisfy the inequality. Show the points on a coordinate grid.

    (c) Place the grid from part (b) on top of the new grid you created in Step C. What does the overlap of these two regions represent? What information does this give Heather?

9. How could you find the "region" given by an inequality without plotting a series of points? Explain the process using Heather's constraints from Investigation 3.

10. Does the region shown in Question 8 (c) resemble the shape you found in Investigation 3, part A? Is this to be expected? Why?

11. Describe how you could find the entire possible "region" without plotting a series of different points. Explain the process using Heather's constraints from Investigation 3.

Think about...

The Constraints
Discuss why the total amount of fabric used to make chairs and couches can be equal to 110 m. Is this reasonable?

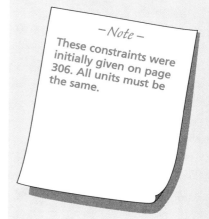

– Note –
These constraints were initially given on page 306. All units must be the same.

Check Your Understanding

Read each of Problems A to E. Then, for each problem, answer Questions 12, 13, 14, and 15 given after Problem E.

A. Manufacturing

A firm manufactures bicycles and tricycles, making a profit of $50 on each bicycle and $30 on each tricycle. Find the maximum profit that can be made under the following conditions:

- the maximum number of frames that can be made each month is 80;
- it takes 2 h to assemble a bicycle;
- it takes 1 h to assemble a tricycle;
- the assembly machine is available for only 100 h each month due to maintenance schedules and time needed to manufacture other products; and
- labour costs are $12.00 per hour.

— Note —
A table can be used to summarize the information in these problems to find possible solutions. The table could be drawn or constructed using spreadsheet software. If you set up the cells appropriately, a spreadsheet will calculate the values in all the cells once you enter the number of chair bundles and the number of couch bundles.

B. Sporting Goods Manufacturing

A sports equipment manufacturer makes basketballs and soccer balls. Two machines are used to make the balls and the time required on each machine is shown.

	Time on Machine A	Time on Machine B
Basketball	2 min	1 min
Soccer Ball	1 min	2 min

Each day, Machine A is available for only 110 min, while Machine B is available for 140 min. If the profit on a basketball is $2.30, and the profit on a soccer ball is $1.95, calculate how many of each type of ball should be made to realize a maximum profit.

— Note —
The difference in time available for Machine A and Machine B is because they are also needed to manufacture baseballs and volleyballs.

C. Hockey Stick Manufacturing

Janet and Sam, two employees of a hockey-stick manufacturer, both make hockey sticks and goalie sticks. Janet can make 3 hockey sticks and 3 goalie sticks per hour, while Sam can make 4 goalie sticks and 1 hockey stick per hour. At least 12 hockey sticks and 30 goalie sticks need to be made each day. If Janet earns $7/h and Sam earns $6/h, how long should each person spend making hockey sticks and making goalie sticks to minimize the cost?

D. Farming

To produce top-quality apples, Dan needs to use 7.3 kg of nutrient A and 4.7 kg of nutrient B for each apple tree each year. Two suppliers sell fertilizers but the amounts of nutrients A and B vary in each brand as shown in the chart. How many kilograms of each brand from each supplier does Dan need to create a fertilizer that provides the required amounts of nutrients A and B at the least cost?

Supplier	Amount of Nutrient A	Amount of Nutrient B	Cost per Kilogram
Erunam	40%	60%	$2.40
Goodwin	90%	10%	$3.00

E. Investments

Mai wants to invest between $15 000 and $20 000 in two mutual funds. One is a balanced investment fund and the other is a more risky high-return fund. She wants at least 60% more money in the balanced fund than the risk fund. However, she wants no more than $10 000 in the balanced fund. The balanced fund is expected to grow by 8% each year. The risk fund is expected to grow by 18% each year. The risk fund fluctuates throughout the year, however, and could be down at a time when she needs to cash it. How much should she put in each fund to maximize the interest she earns?

12. List all of the constraints in the problem. Write each constraint using mathematical symbols.

13. Find a solution to each of the problems using guess and check. Describe any difficulties you had doing this. Write a profit or cost relation for each.

14. Rate the confidence you have in your solution using the scale below. Give reasons why you picked the value that you did.
    0 = no confidence
    3 = some confidence, but have a feeling the solution is not accurate
    5 = fairly confident
    8 = quite confident but not 100% positive there isn't a better solution
    10 = completely confident with no doubt that optimal solution found

15. List any pieces of extraneous information provided in each problem.

> **Did You Know?**
> Millions of dollars are invested in mutual funds each year. Most of this investment money comes from Registered Retirement Savings Plans (RRSPs).

extraneous information – information that does not help you find a solution

Think about...

Question 15
How can you determine if information is extraneous?

16. Write an inequality to represent each situation below. Graph the inequality.
    (a) Ostrich farming has become popular and lucrative in Manitoba. One ostrich can lay up to 70 eggs in one year.
    (b) For many farmers in the Tai Lake Valley of China, silk production provides the main source of income. To form a cocoon, each silk worm produces a single strand up to 1.5 km long. These threads are used to make silk fabric.

17. Jorge must write a term paper on the writing of Canadian author Margaret Laurence. His bibliography must include at least 10 sources. At least three sources must be original works by Margaret Laurence and the rest can be secondary sources.
    (a) Write inequalities to represent the number of original sources, p, and the number of secondary sources, s, that Jorge must use.
    (b) Graph the inequalities on the same coordinate grid.

18. An international telephone call from Canada to China costs between $1.56 per minute and $5.58 per minute, depending on the time of day, for the first minute. Each additional minute costs 95% as much as the first minute.
    (a) Write the two inequalities that represent the relationship between the cost of the first minute and the cost of each additional minute.
    (b) Graph the inequalities.

— Note —
For Question 18, you will have to select your own variables to write the inequalities.

"What If ...?"

Heather and her brother Colin both looked at the graphs of the constraints that had been drawn for Heather's job. Heather wanted Colin's input on whether the job would be profitable.

Heather said, "You know, each of the inequalities we graphed had either ≤ or ≥ in them and solid lines in the graph. What if we graphed $5x + 3y < 110$ instead of $5x + 3y \leq 110$? What would this graph look like?"

Chapter 7 Linear Programming

Colin thought for a minute and then responded, "That's an interesting question. A solid line shows that the points on the line $5x + 3y = 110$ are included in the solution. If you used the inequality $5x + 3y < 110$, the points on the line $5x + 3y = 110$ would not be included as part of the region. Showing the same graph with a dotted line instead of a solid line indicates that the points on the line are not included. Let me show you."

> **– Note –**
> Remember: When multiplying or dividing both sides of an inequality by a negative number, the direction of the inequality changes. Why do you think this is so?
>
> For example:
> if $3 > 1$, then $-3 < -1$.
> if $-x < 3$, then $x > -3$.
> if $x + y < 3$, then $-x - y > -3$.

$5x + 3y \leq 110$

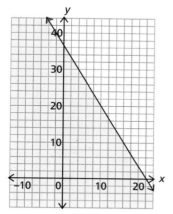

$5x + 3y < 110$

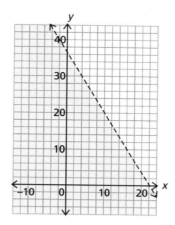

Check Your Understanding

19. Sketch the feasible region that corresponds to each set of inequalities.

(a) $y \leq 6 - x$ and $y \geq x - 2$
(b) $x + y > 9$ and $x - y \leq 3$
(c) $3x - y \leq -2$ and $x - y \leq 6$
(d) $3y - 2x + 12 \geq 0$ and $3y - x \leq 0$
(e) $3x - y < 4$ and $x - 2y > 3$
(f) $3x + 2y < 6$ and $x \geq 0$ and $y \geq 0$
(g) $x + y \leq 9$ and $x - y \leq 3$ and $x \geq 0$ and $y \geq 0$

feasible region – a shaded region on a graph indicating that all points within the region are possible solutions to the problem

Think about...

Question 20

Why are the graphs of the inequalities shaded in the first quadrant only? The quadrants are reviewed below.

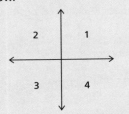

CHALLENGE yourself

Suppose you were the manager of a local clothing store. You have a radio advertising budget equal to ten times the cost of the most expensive 60-s advertisement. You can run, at most, 22 ads per day. How many ads will you run in the morning and how many will you run in the afternoon so that your ads will be heard by the greatest number of people? Contact a local radio station in your area. Find out:

- the advertising rates for the morning and the afternoon; and
- the number of listeners, on average at any one time, for the morning and the afternoon.

20. Suppose the following constraints were given for Heather's problem. All other constraints would stay the same. Write each as an inequality. Graph the constraints on one graph.

    (a) At least 9 couches are needed.
    (b) At least 11 chairs are needed.
    (c) There are 150 m of fabric available.
    (d) Heather can work a minimum of 25 h per week.

21. Write an inequality to represent each of the following.

    (a)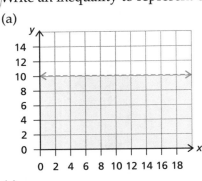

    (b)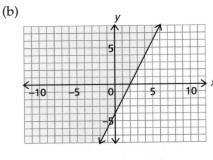

    (c)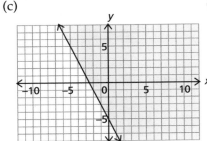

    (d)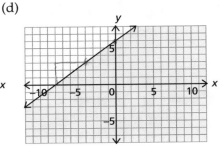

22. (a) Write the inequalities that define this region.

    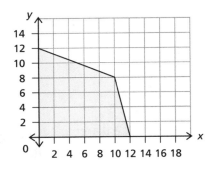

    (b) Create a context of your own for which the graph above could represent the feasible region.

7.3 Connecting the Region and the Solution

Graphing the feasible region allows you to test possible solutions to a problem. In this section, you will develop a procedure that will allow you to find the **optimal solution** to a problem more directly and exactly.

As you complete Investigation 4, remember that you wrote Heather's income function, usually called the **objective function**, as $I = 5.5x + 13.5y$, in which x represents the number of chairs, y represents the number of couches, and I represents Heather's total income.

optimal solution – the solution that best meets the constraints in the problem

objective function – a function that allows you to find the maximum or minimum values using the given constraints

Investigation 4
Finding Heather's Exact Maximum Income

Purpose
To investigate how to accurately calculate Heather's maximum potential income.

Procedure
The feasible region for Heather's potential income is shown below.

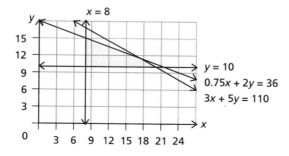

A. Suppose Heather earned $209 over a two-week period. Find at least five ordered pairs in the feasible region that give this income. Mark them on the graph.

B. Share your ordered pairs from Step A with other students. Add their ordered pairs to your graph. Describe the pattern shown by all of the ordered pairs.

C. Repeat Steps A and B for an income of $220 and then for an income of $190. Compare the patterns formed by the ordered pairs in each case. Are these patterns helping you to get closer to a solution to Heather's problem?

Think about...

Step A
Why can Heather have ordered pairs that do not contain whole numbers?

Investigation Questions

1. Based on your results, which of the following statements describes where Heather's maximum income will fall? Give reasons for your choice.
   (a) less than $190
   (b) greater than $190 but less than $209
   (c) greater than $209 but less than $220
   (d) greater than $220

2. The graph of Heather's region is shown below. A profit line has been drawn in colour on the graph.

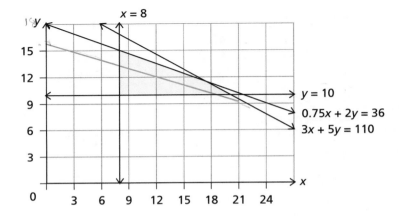

   (a) Verify that $212 = 5.5x + 13.5y$ is the equation of this line.
   (b) Does this line show Heather's maximum income? Why?
   (c) Suppose Heather predicted she would earn $222. Show the profit line on your graph.
   (d) Predict the maximum income Heather can earn. How did you make your prediction?

3. Find Heather's maximum income.
   (a) Estimate the point on your graph that represents Heather's maximum income. Estimate Heather's maximum income. Why did you choose this point?
   (b) Why can you discard the rest of the graph and use only the feasible region to identify the point where Heather's maximum income will occur?
   (c) Use the scale at the left to describe your confidence in your solution. Justify your choice.

CHALLENGE yourself

Suppose the rate of pay for cutting couch bundles were changed to $12 per bundle. How would this change Heather's maximum income? How would it change the objective function?

0 = no confidence

3 = some confidence, but have a feeling the solution is not accurate

5 = fairly confident

8 = quite confident but not 100% positive there isn't a better solution

10 = completely confident with no doubt that optimal solution found

4. Review the process you used to accurately find Heather's maximum income potential. In your own words, describe the process. Use examples from Heather's problem to support your description.

5. As her income potential increases, what happens to the slope of the objective function? What happens to the intercepts?

6. Suppose Heather earned $13.50 for each couch bundle cut and $13.50 for each chair bundle cut. All other constraints stay the same. How will this affect the point at which Heather receives the maximum amount of money? Sketch the graph to help answer this question.

Check Your Understanding

7. For the graph shown below, three objective functions are given.
   - Find the equation of each objective function for the graph.
   - What do you notice about them?
   - What will be the profit for each objective function? What is the maximum profit? How do you know?
   - Do your answers change if the function asks for expenses instead?

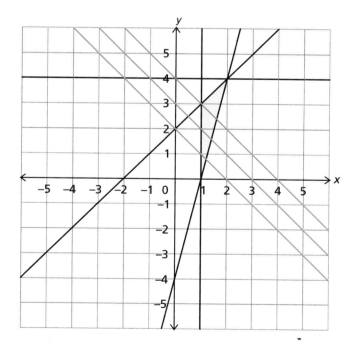

Focus D: Systems of Equations

system of equations –
two or more equations involving the same variable quantities

In your work on Investigation 4, knowing how to find the point where two lines intersect helped you determine the exact point showing Heather's maximum income. The following example shows one method for finding the coordinates of the intersection point of two lines.

Example 1
Find the point where the following two lines intersect. (Finding the point where the two lines intersect is called **solving a system of equations**.)

$y = 4x - 1$ ①
$y = x + 3$ ②

The Solution
How do the values you have found for *x* and *y* relate to the graph that shows $y = 4x - 1$ and $y = x + 3$? Why might you also need the value of *y* in Heather's situation?

Solution

In Chapter 3, when you graphed equations like $y = 4x - 1$ and $y = x + 3$ on the same coordinate grid, the *x*-coordinate of the intersection point was the solution to $4x - 1 = x + 3$ as shown below.

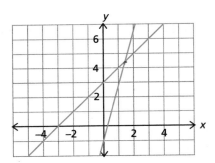

You can also use the skills developed in Chapter 3 to solve for *x* and *y*.

$4x - 1 = x + 3$
$ 3x = 4$
$ x = \dfrac{4}{3}$

When solving a system of equations, you also need to find the value of *y*. Take the value of *x* that you have just found and substitute it into one of the original equations.

$y = x + 3$
$= \frac{4}{3} + 3$
$= \frac{13}{3}$

The point of intersection is $(\frac{4}{3}, \frac{13}{3})$.

Focus Questions

8. (a) A dog-food manufacturer sells both a high-fibre dog food and a low-fat dog food. The profit for the company can be represented by $P = 3x + 2.50y$, where x represents the mass of a can of the high-fibre dog food and y represents the mass of a can of the low-fat dog food. Each graph below represents a different set of manufacturing constraints. Find the maximum profit for each set of constraints. Check all answers algebraically.

(i)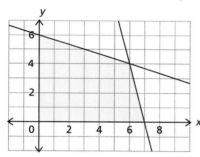

(ii)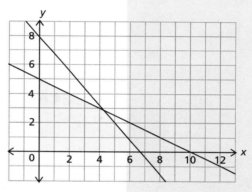

(iii)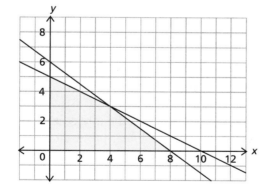

(b) Refer to the graph in part (i). Change the objective function so that the maximum profit can be found at (0, 6).

9. (a) Refer back to the graphs you used in Investigation 4 when finding Heather's maximum potential income. Find the coordinates of each vertex of the feasible region.
   (b) Use these vertices, and the income function, to find Heather's income at each vertex.
   (c) What is the maximum income that Heather can earn?

10. If there is only one possible solution for a system of equations, it is called a unique solution.
   (a) Graph two straight lines that have a unique solution. What is true about the slopes of the lines?
   (b) Graph two straight lines that will never intersect. This will represent a system of equations with no solution. What is true about the slopes of the lines?
   (c) Graph two straight lines that intersect in two or more points. This represents a system of equations with many solutions. What do you notice about the lines? What do you notice about the slopes of the lines?

CHALLENGE yourself

For each of parts (a), (b), and (c) in Question 10, write a situation in which each case might arise.

Check Your Understanding

11. Refer to your method of solving systems of equations in Focus D on the previous page. Use the same method to solve the following systems of equations.

   (a) $y = 2x - 1$
   $y = 4x - 3$

   (b) $4x = -y + 5$
   $2y = -4x - 22$

   (c) $y = 0.5x + 2.5$
   $6.5 = 3.5x - y$

   (d) $2y - 6x = 1$
   $y = \frac{1}{4}x + 4$

   (e) $y = \frac{2}{3}x + \frac{1}{2}$
   $y = \frac{4}{5}x + \frac{3}{4}$

   (f) $y = 2(x - 5)$
   $\frac{y}{-4} = 3 - x$

— Note —
You first explored how to rewrite equations and formulas in Chapter 3.

12. Graph the following regions and find the intersection points of the border for each region. Since in all linear programming problems the numbers are real, all regions must be defined for $x \geq 0$ and $y \geq 0$. Find the greatest value for the objective function given by $I = 2x + 12y$.

   (a) $y \leq 4x - 1$
   $y \leq -2x + 7$

   (b) $y \geq x$
   $y \leq 5x - 4$

   (c) $y \leq 2x + 1$
   $y \geq x + 3$

FOCUS E: Solve Systems of Equations

Henri manages a manufacturing plant that makes softballs and hard baseballs. Each hard ball requires 1 min on the stitching machine and 3 min on the covering machine. Each softball requires 2 min on each machine. The stitching machine is available only 100 min each day, while each ball-covering machine is available for 3 h each day. The profit on a hard ball is $2. The profit on a softball is $3. How many of each should be manufactured each day?

To find the number of each, Henri decided to set up a system of equations because an estimate was not accurate enough for his needs.

Let x represent the number of hard balls.
Let y represent the number of softballs.
$x + 2y \leq 100$ ① $3x + 2y \leq 180$ ②

Henri decided to rewrite equation ② to solve for y.
$y = 90 - 1.5x$ ③

Since this equation came from equation ②, you need to substitute it into equation ① and solve for x.
$x + 2(90 - 1.5x) = 100$
$x + 180 - 3x = 100$
$x = 40$

Now find y by substituting $x = 40$ into equation ③.
$y = 90 - 1.5x$
$y = 90 - 1.5(40)$
$y = 30$

The intersection point is $(40, 30)$, which means that Henri will manufacture 40 hard balls and 30 softballs.

Focus Questions

13. Find any other intersection points from Henri's problem. (Hint: Think of other constraints.)

14. Find Henri's maximum profit. Do you need to graph the feasible region first? Why?

Think about...

Henri's Decision

Could Henri have rewritten equation ① to solve for x instead of rewriting equation ② to solve for y? Which is the better choice? Why?

Think about...

The Substitution

Try substituting back into equation ②. Why is substituting into equation ① necessary?

— Note —
The intersection point $(40, 30)$ gives only one possible solution. Other solutions need to be found.

Check Your Understanding

Think about...

Question 15
Use the points of intersection for Heather's problem. Why does substituting these points into the profit function give you the optimal solution for Heather?

15. Find the point of intersection of all boundary lines in Heather's problem using Henri's method. Are your answers the same? Is this to be expected? What is Heather's maximum income?

16. Jane makes macramé jewelry. The number of bracelets she makes is represented by the first b and the number of necklaces she makes is represented by n. Find the number of necklaces and number of bracelets made each week in each of the following systems. Are there any answers that are unreasonable? Explain by solving each of the following systems.

(a) $b + 2n = 3$
    $b - n = 3$

(b) $2b + n = 6$
    $4b - n = 9$

(c) $2b + 4n = 5$
    $b + n = 2$

(d) $3b + n = 12$
    $2b + 3n = 30$

(e) $b = 5n + 8$
    $b = 2n - 1$

(f) $b - 2n = 3$
    $b + n = 6$

(g) $4b = 3 + 15n$
    $3b + 2n = 21$

(h) $2b + 7 = n$
    $2b = -3n + 8$

(i) $8 + 3n = 3b$
    $5b + 12 = -7n$

17. Solve each of the following systems of equations.

(a) $\frac{1}{3}x - y = -2$
    $x - \frac{2}{3}y = 1$

(b) $y + \frac{3}{4}x = 4$
    $x - \frac{1}{4}y = -1$

(c) $\frac{1}{2}x + 2y = 4$
    $x - y = 2$

(d) $2a - \frac{1}{2}b = 4$
    $\frac{2}{3}a = b - 2$

(e) $\frac{1}{2}p + \frac{1}{3}q = 5$
    $\frac{1}{3}q - p = -4$

(f) $\frac{1}{2}m + 3n = 2$
    $\frac{2}{3}m + n = 5$

18. In each of the following systems of equations, x represents the number of video games played per hour and y represents the number of pinball games played per hour at a local video arcade. Find the number of video games and the number of pinball games played per hour in each situation. Are there any answers that are unreasonable? Explain.

(a) $1.9x - 3.3 = 2.8y$ and $4.1y + 5.2x - 8.3 = 0$
(b) $2.4x = 2.2y - 3.2$ and $1.6x + 1.2y = 3.2$
(c) $1.25x - 0.25y = 0.75$ and $0.45y = 0.65x + 1.35$
(d) $3(x - 1) - 2(y + 1) = -14$ and $3x - y = -6$
(e) $2.3(1.1x - 3.2y) = 3.6$ and $2.4y + 1.8x = 4.3$
(f) $3(x + 2y) = 48$ and $\frac{1}{4}(x - 3y) = \frac{1}{2}(x - 5)$
(g) $\frac{1}{4}(x - y) = 48$ and $\frac{3}{5}(x + 7y) = 12$
(h) $\frac{2}{3}(2x - 3y) = \frac{3}{4}$ and $\frac{3}{4}(y + 3x) = \frac{1}{2}$

326 Chapter 7 *Linear Programming*

19. (a) Sketch the region defined by each of the following pairs of inequalities. Find the maximum profit if the objective (profit) function is given by $P = 3x + 2y$. For each, the values of x and y are greater than zero.

   (i) $y \leq 4 - 2x$ and $y \geq 1$
   (ii) $x - 2y \geq -12$ and $3x - y \leq 5$
   (iii) $2x + 3y \leq 6$ and $3x + 2y \geq 6$

   (b) How can the slope of the objective function help you determine at what vertex the maximum profit will occur?

The problems that follow are taken from Section 7.2 on page 314.

20. Solve each problem using the method for finding the point of intersection that you have investigated in this section.

21. Compare your results now with any results you found earlier. Describe how they compare. Explain why this happens.

A. Manufacturing

A firm manufactures bicycles and tricycles, making a profit of $50 on each bicycle and $30 on each tricycle. Find the maximum profit that can be made under the following conditions:

- the maximum number of frames that can be made each month is 80;
- it takes 2 h to assemble a bicycle;
- it takes 1 h to assemble a tricycle;
- the assembly machine is available 100 h each month; and
- each requires 2 h of labour at $12.00 per hour.

B. Manufacturing Sporting Goods

A sports equipment manufacturer makes basketballs and soccer balls. Two machines are used to make the balls and the time required on each machine is shown.

	Time on Machine A	Time on Machine B
Basketball	2 min	1 min
Soccer Ball	1 min	2 min

Each day, Machine A is available for only 110 min, while Machine B is available for 140 min. If the profit on a basketball is $2.30, and the profit on a soccer ball is $1.95, calculate how many of each type of ball should be made to realize a maximum profit. (Assume the company sells all of the balls it manufactures.)

Think about...

Each Solution
Where does the optimal solution to any linear programming problem appear to be?

C. **Hockey Stick Manufacturing**

Janet and Sam, two employees of a hockey-stick manufacturer, both make hockey sticks and goalie sticks. Janet can make 3 hockey sticks and 3 goalie sticks per hour, while Sam can make 4 goalie sticks and 1 hockey stick per hour. At least 12 hockey sticks and 30 goalie sticks need to be made each day. If Janet earns $7/h and Sam earns $6/h, then how long should each person spend making hockey sticks and making goalie sticks to minimize the cost?

D. **Farming**

To produce top-quality apples, Dan needs to use 7.3 kg of nutrient A and 4.7 kg of nutrient B for each apple tree each year. Two suppliers sell fertilizers but the amounts of nutrients A and B vary in each brand as shown in the chart. How many kilograms of each brand from each supplier does Dan need to create a fertilizer that provides the required amount of nutrients A and B at the least cost?

Supplier	Amount of Nutrient A	Amount of Nutrient B	Cost per Kilogram
Erunam	40%	60%	$2.40
Goodwin	90%	10%	$3.00

E. **Investments**

Mai wants to invest between $15 000 and $20 000 in two mutual funds. One is a balanced investment fund and the other is a more risky high-return fund. She wants at least 60% more money in the balanced fund than the risk fund. However, she wants no more than $10 000 in the balanced fund. The balanced fund is expected to grow by 8% each year. The risk fund is expected to grow by 18% each year. The risk fund fluctuates throughout the year, however, and could be down if she needs to cash it in an emergency. How much should she put in each fund to maximize the interest she earns?

PUTTING IT TOGETHER

PROBLEM SOLVING

A Radio Show

As the sales manager for CTPN radio station, you want to sell exclusive advertising rights on a half-hour music program called "Martina's Magic." Music will be played for x min, commercials will run for y min, and Martina will talk the rest of the time.

You want to find the maximum number of listeners you can expect to have at any time throughout the show because you know that potential advertising customers will need this information. Through research, you determine that the total number of listeners, N, is influenced as follows:

- the audience increases by 500 for every minute music is played;
- the audience decreases by 500 for every minute Martina speaks;
- to be profitable, the program must run at least 5 min of commercials. Government regulations state that a maximum of 12 min of commercial time be used per half hour; and
- the amount of time commercials are played must not exceed the amount of time in which music is played.

(a) Write inequalities to represent the constraints. Let x represent the time length of a song and y represent the time length of a commercial.

(b) Graph the region represented in (a).

(c) Write an equation for the total number of listeners as $N = $ ....

(d) What is the maximum number of listeners for the show?

EXTENSION

A Radio Show

Another important consideration is the cost to produce the show. The sales manager offers the following costs to minimize the amount of time the client needs to spend making a decision:

- royalty payments of $400 per minute are required;
- commercials cost $200 per minute to make and run; and
- each minute Martina speaks costs $300/min.

(a) Write an equation in the form $C = \boxed{}$ that can be used to calculate the cost to produce the show.

(b) What is the minimum cost for the show?

(c) What is the minimum cost to attract the maximum number of listeners?

(d) What is the maximum cost for attracting the maximum number of listeners?

CASE STUDY

The Diet Problem

www.mcs.anl.gov/home/otc/Guide/Case Studies/diet

The goal of the "Diet Problem" is to find the cheapest combination of foods that will satisfy all the daily nutritional requirements of a person. The objective is to minimize cost and meet constraints which require that nutritional needs be satisfied. You must include constraints that regulate the number of calories and amounts of vitamins, minerals, fats, sodium, and cholesterol in the diet.

The diet problem was motivated by the American Army's desire to meet nutritional requirements of soldiers at a minimum cost. The initial solution was found in 1939 by 9 people working 120 days each.

When doing the problem:

- choose acceptable foods for your menu; and
- try to create an optimally low-cost menu that meets the constraints.

This problem can be explored on the Internet using built-in computer programs that allow you to solve the problem using constraints you impose. Try it! You may be interested to discover how people select a nutritionally balanced diet. Remember, however, that what is mathematically acceptable may not be practical, possible, or real.

Have fun with this case study!

— Note —
Remember that these results are for information only. The optimal diet may not be recommended by your doctor.

REVIEW

Key Terms

	page
constraint	306
extraneous information	315
feasible points	311
feasible region	317
feasible solution	311
graphical solutions	319
inequality	312
intersection point	322
linear programming	305
objective function	320
optimal solution	319
system of equations	322

You Will Be Expected To

- identify the variables in a problem and decide upon constraints given in the problem.
- graph linear inequalities in two variables to define a feasible region.
- identify a quantity to be maximized or minimized, and express the quantity in terms of the variables. Form the objective function from known parameters in a problem.
- develop and practise methods for solving a system of linear equations.
- formalize the relationship between the feasible region, the intersection points on the border of the feasible region, and the optimal solution to a problem.

Summary of Key Concepts

Vince is a company's sales representative for Newfoundland and Labrador (NF) and New Brunswick (NB). Since he must pay his own expenses, he reviews his expenses from the last few years to develop a budget. He decides upon an "average" daily travel budget of $100 per day when traveling in NF and an "average" daily travel budget of $120 when traveling in NB. To be profitable, he needs to stay within an annual travel budget of $18 000. As part of his contract, he must spend at least 50 d in NF and 60 d in NB.

Vince also has leased a car. In NF, he travels an average of 275 km per day. In NB, he travels an average of 110 km per day. His lease lets him travel a maximum of 26 400 km per year.

Historically, sales average $3000 per day in NF and $2800 per day in NB. What are the maximum sales Vince can expect? How many days should he travel in each province?

7.1 Exploring an Optimization Problem

Write inequalities to represent the constraints in this problem.

The constraints can be classified under these three headings.

Constraints on days in each province

$n \geq 50 \qquad p \geq 60$

Constraints on daily travel budget

$100n + 120p \leq 18\,000$

Constraints on yearly travel

$275n + 110p \leq 26\,400$

— Note —
For this problem, p represents number of days traveling in New Brunswick and n represents number of days traveling in Newfoundland and Labrador.

7.2 Exploring Possible Solutions

You can also show the constraints on a graph and place appropriate shading for the feasible region. For example, to decide where to shade, you can select a test point on one side of the line and test the point to determine if it belongs in the region.

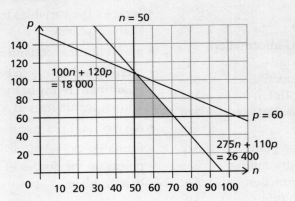

Historically, sales average $3000 per day in NF and $2800 per day in NB. The objective function, P, can be written as $P = 3000n + 2800p$.

7.3 Connecting the Region and the Solution

The objective function is always drawn through the feasible region. Evaluating parallel objective functions will allow you to determine the maximum profit at minimum cost. To simplify the process, find the points that make up the vertices of the border on the feasible region. In this case, there are four different systems that need to be solved in order to do so.

System 1

$$n = 50$$
$$p = 60$$

The solution is $n = 50$ and $p = 60$.

System 2

$$100n + 120p = 18\ 000$$
$$n = 50$$
$$100n + 120p = 18\ 000$$
$$100(50) + 120p = 18\ 000$$
$$120p = 13\ 000$$
$$p = 108.333\ldots$$

Since you already have the value of n, you can substitute it directly into the equation.

The solution is $n = 50$ and $p = 108.333\ldots$

System 3

$$275n + 110p = 26\,400$$
$$p = 60$$

$$275n + 110p = 26\,400$$
$$275n + 110(60) = 26\,400$$
$$275n = 19\,800$$
$$n = 72$$

The solution is $n = 72$ and $p = 60$.

System 4

$$100n + 120p = 18\,000 \quad ①$$
$$275n + 110p = 26\,400 \quad ②$$

From ①

$$100n = 18\,000 - 120p$$
$$n = 180 - 1.2p \quad ③$$

Substitute ③ into ①.

$$275(180 - 1.2p) + 110p = 26\,400 \qquad 100n + 120(105) = 18\,000$$
$$49\,500 - 330p + 110p = 26\,400 \qquad 100n + 12\,600 = 18\,000$$
$$49\,500 - 220p = 26\,400 \qquad\qquad 100n = 5400$$
$$-220p = -23\,100 \qquad\qquad\qquad n = 54$$
$$p = 105$$

Thus, $p = 105$ and $n = 54$.

> **– Note –**
> When you graph a feasible region, there are different lines that make up the border of the region. The vertices are the points where any two lines intersect. You can use your skills with solving systems of equations to find these vertices.

Check the coordinates of the points found in the profit function to find the maximum profit. What is the maximum profit Vince can make? How many days should he travel in each province?

Value of n	Value of p	$P = 3000n + 2800p$
50	60	$318 000
54	105	$456 000
50	108.333	$453 333.33
72	60	$384 000

Thus, Vince should travel 54 d in NF and 105 d in NB to maximize sales at $456 000.

PRACTICE

7.1 Exploring an Optimization Problem

For each problem:
- decide whether you have enough information to solve the problem as stated; and
- if a problem can be solved, solve it. Otherwise, explain why it can't be solved. If you had to make any assumptions when solving the problem, list them.

1. To manufacture cushions and pillows, a firm uses two machines, A and B. The time required on each machine is shown. Machine A is available for one full shift of 9 hours each day and Machine B is available for parts of two shifts for a total of $10\frac{2}{3}$ h each day.

	Time on Machine A	Time on Machine B
Pillows	2 min	1 min
Cushions	1 min	2 min

Find the mix of cushions and pillows that will earn the manufacturer the maximum profit.

2. A firm manufactures two-bulb bedroom lamps and four-bulb living room lamps. Determine how many of each lamp they should manufacture to yield a maximum profit. Profit on a two-bulb lamp is $20 and profit on a four-bulb lamp is $35.

3. It is recommended that the daily cattle diet of grass and hay be supplemented with 19 g of iron and 12 g of riboflavin. Two cattle supplements are available which contain both nutrients in different amounts as shown in the table.

Feed	Iron	Riboflavin
Husky	5%	2%
Vibrant	2%	3%

Husky feed sells for $25/kg and Vibrant feed sells for $32/kg.
(a) How many kilograms of each feed are required to feed 100 cattle as economically as possible?
(b) What is the total cost to feed the cattle?

4. The Zimmer Watch Company manufactures two types of watches — a watch with hands (regular) and a watch with a digital display. The digital model requires 1.5 h of machine time and 1 h of jeweler time. The regular model requires 30 min of machine time and 2 h of jeweler time. The profit on a digital model is $25 and on a regular model is $18. How many of each type should be manufactured daily to achieve maximum profits?

5. A furniture manufacturer makes desks and computer tables using a combination of pine and oak. Each desk takes 5 h to make, while a computer table takes 4 h. The manufacturer has guaranteed sales for everything that can be produced. The manufacturer has a 150-m sheet of oak and a 100-m sheet of pine. There are 90 h for manufacturing. How many of each should be made to maximize profits?

7.2 Exploring Possible Solutions

6. Graph the following regions.
 (a) $y \geq 2x - 4$
     $2x - y > 6$
 (b) $x - 3y \leq 6$
     $3x - y \geq 3$
 (c) $5x + 2y \leq 10$
     $2x + 5y \geq 10$
 (d) $3x + 2y \geq 12$
     $3x + 5y \geq 30$
 (e) $2x - y \geq 3$
     $x + y \leq 3$
 (f) $y - x \leq 4$
     $x + y > 1$

7. Suppose the following constraints are added to the pillow manufacturing problem of Question 1.
   - The profit on a cushion is $3.20.
   - The profit on a pillow is $1.20.
   (a) Write the constraints as inequalities.
   (b) Decide what other constraints there are that are not explained in the problem. Write them as inequalities.
   (c) Graph all the inequalities on one graph.

8. Repeat parts (a), (b), and (c) of Question 7 by adding the following information to the constraints of Question 2.
   - Each day the manufacturer receives 480 bulbs.
   - Each day the manufacturer receives 180 lamp shades.

9. Find the inequalities that are represented on each graph. Are there other restrictions shown by the graph?

   (a)
   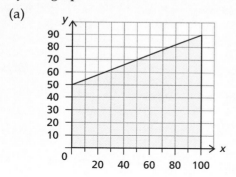

   (b)
   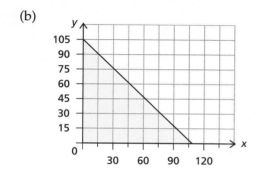

7.3 Connecting the Region and the Solution

10. Casey Sports manufactures golf balls and tennis balls. The number of golf balls manufactured each day is represented by x in each pair of equations below. The number of tennis balls manufactured is represented by y. Find the point of intersection for each of the following. What does the point of intersection represent?

    (a) $x = 6 - 4y$
    $2x - y = 3$

    (b) $2x - 3y = 13$
    $4x + y = 5$

    (c) $3x - y = 3$
    $4x + 2y = 14$

    (d) $x - 5y = 10$
    $3x - y = 16$

    (e) $y = -2x + 2$
    $y = 0.5x - 3$

    (f) $y = 3x - 4$
    $y = 0.5x - 1$

11. Burlington Runners repairs sports shoes, particularly tennis shoes and jogging shoes. Two operators perform different functions to repair each shoe and the times required for each operation are given below.
    - Tennis shoe: 16 min to strip and 12 min to re-sew
    - Jogging shoe: 8 min to strip and 16 min to re-sew

    The profits on a tennis-shoe repair are $3.00 and on a jogging-shoe repair are $5.00. Two people repair shoes, working eight hours each day. How many pairs of each type of shoe should ideally be repaired daily to maximize profits?

12. Jack and Carlene are making juice for a mall kiosk. They use orange juice and fruit drink in their punch. During the summer, they can sell between 12 L and 15 L per day. Their recipe calls for them to have at least twice as much orange juice as fruit drink in the final punch. They estimate that orange juice will cost them $2.49 per litre and fruit drink will cost them $1.53 per litre. How much of each type should they buy to minimize the cost of the punch?

Trigonometric Tables

Angle in Degrees	Sine	Cosine	Tangent	Angle in Degrees	Sine	Cosine	Tangent
0	0.0000	1.0000	0.0000	45	0.7071	0.7071	1.0000
1	0.0175	0.9998	0.0175	46	0.7193	0.6947	1.0355
2	0.0349	0.0994	0.0349	47	0.7314	0.6820	1.0724
3	0.0523	0.9986	0.0524	48	0.7431	0.6691	1.1106
4	0.0698	0.9976	0.0699	49	0.7547	0.6561	1.1504
5	0.0872	0.9962	0.0875	50	0.7660	0.6428	1.1918
6	0.1045	0.9945	0.1051	51	0.7771	0.6293	1.2349
7	0.1219	0.9925	0.1228	52	0.7880	0.6157	1.2799
8	0.1392	0.9903	0.1405	53	0.7986	0.6018	1.3270
9	0.1564	0.9877	0.1584	54	0.8090	0.5878	1.3764
10	0.1736	0.9848	0.1763	55	0.8192	0.5736	1.4281
11	0.1908	0.9816	0.1944	56	0.8290	0.5592	1.4826
12	0.2079	0.9781	0.2126	57	0.8387	0.5446	1.5399
13	0.2250	0.9744	0.2309	58	0.8480	0.5299	1.6003
14	0.2419	0.9703	0.2493	59	0.8572	0.5150	1.6643
15	0.2588	0.9659	0.2679	60	0.8660	0.5000	1.7321
16	0.2756	0.9613	0.2867	61	0.8746	0.4848	1.8040
17	0.2924	0.9563	0.3057	62	0.8829	0.4695	1.8807
18	0.3090	0.9511	0.3249	63	0.8910	0.4540	1.9626
19	0.3256	0.9455	0.3443	64	0.8988	0.4384	2.0503
20	0.3420	0.9397	0.3640	65	0.9063	0.4226	2.1445
21	0.3584	0.9336	0.3839	66	0.9135	0.4067	2.2460
22	0.3746	0.9272	0.4040	67	0.9205	0.3907	2.3559
23	0.3907	0.9205	0.4245	68	0.9272	0.3746	2.4751
24	0.4067	0.9135	0.4425	69	0.9336	0.3584	2.6051
25	0.4226	0.9063	0.4663	70	0.9397	0.3420	2.7475
26	0.4384	0.8988	0.4877	71	0.9455	0.3256	2.9024
27	0.4540	0.8910	0.5095	72	0.9511	0.3090	3.0777
28	0.4695	0.8829	0.5317	73	0.9563	0.2924	3.2709
29	0.4848	0.8746	0.5543	74	0.9613	0.2756	3.4874
30	0.5000	0.8660	0.5774	75	0.9659	0.2588	3.7321
31	0.5150	0.8572	0.6009	76	0.9703	0.2419	4.0108
32	0.5299	0.8480	0.6249	77	0.9744	0.2250	4.3315
33	0.5446	0.8387	0.6494	78	0.9781	0.2079	4.7046
34	0.5592	0.8290	0.6745	79	0.9816	0.1908	5.1446
35	0.5736	0.8192	0.7002	80	0.9848	0.1736	5.6713
36	0.5878	0.8090	0.7265	81	0.9877	0.1564	6.3138
37	0.6018	0.7986	0.7536	82	0.9903	0.1392	7.1145
38	0.6157	0.7880	0.7813	83	0.9925	0.1219	8.1443
39	0.6293	0.7771	0.8098	84	0.9945	0.1045	9.5144
40	0.6428	0.7660	0.8391	85	0.9962	0.0872	11.4301
41	0.6561	0.7547	0.8693	86	0.9976	0.0698	14.3007
42	0.6691	0.7431	0.9004	87	0.9986	0.0523	19.0811
43	0.6820	0.7314	0.9325	88	0.9994	0.0349	28.6363
44	0.6947	0.7193	0.9657	89	0.9998	0.0175	57.2900
				90	1.0000	0.0000	undefined

Index

absolute value, 183
accuracy, 8
acute triangle, 293
adjacency matrix, 68
adjacent side, 214
altitude, 294
angle bisector, 293
angle of elevation, 244
apothem, 267
average, 14

bearing, 217
bin, 22
box-and-whisker plot, 18
broken-line graph, 157

capacity, 262
centre of gravity, 295
circumcentre, 292
circumscribed circle, 292
coincident, 294
cone, 259
constraint, 306
continuous, 97
controlled variables, 4
correlation, 192
correlation coefficient, 196
cos X, 236
cube, 271
cube root, 278
cylinder, 259

dependent variable, 2
digraph, 63
dimensions, 68
discrete, 97
dispersion, 27
distribution, 17
distributive property, 113
domain, 97

economy rate, 275

edge, 57
element, 68
equation, 102
equilateral triangle, 264
Eulerian, 58
even vertex, 58
exponential relation, 142
exterior angle, 265
extraneous information, 315
extrapolate, 42

factoring, 135
factors, 135
feasible points, 311
feasible region, 317
feasible solution, 311
frequency polygon, 33
frequency table, 22
function, 165

graph, 56
graphical solutions, 319

hendecagon, 269
heptagon, 264
hexagon, 264
histogram, 22
hypotenuse, 214

identity, 113
incentre, 292
independent variable, 2
index, 279
inscribed circle, 292
interior angle, 265
interpolate, 42
intersection point, 111, 322
intersection (overlap) region, 319

lateral face, 270
lateral surface, 270

like terms, 112
line of best fit, 40
line symmetry, 265
linear, 100
linear inequality, 317
linear programming, 305
lower extreme, 18

major diagonal, 71
mapping notation, 175
matrix, 67
mean, 14
median, 14, 293
median-median line, 192
mode, 14
modeling, 96

network, 56
network graph, 57
nonagon, 267
normal distribution, 36

objective function, 320
obtuse triangle, 294
octagon, 264
odd vertex, 5
opposite side, 214
optimal solution, 319
optimization problem, 306
order of rotational symmetry, 265
outlier, 15

pentagon, 264
perfect cube, 279
perfect square, 278
perpendicular bisector, 293
polygon, 264
precision, 8
principal square root, 226
prism, 259
product matrix, 77
pyramid, 259
pyramid height, 259
Pythagorean theorem, 220

quadratic, 134
quadrilateral, 266
quartile, 19

radicand, 279
random, 33
range, 17, 97
rate, 275
regression, 195
regular polygon, 264
resultant vector, 246
rigidity, 292
rotational symmetry, 265

scale factor, 285
scalene triangle, 293
scatter plot, 40
sequence, 96
similar, 285
sin X, 236
slant height, 270
slope, 117
slope y-intercept form, 117
sphere, 259
square, 264
square matrix, 79
square root, 278
standard deviation, 27
stem-and-leaf plot, 17
summary point, 192
surface area, 270
system of equations, 322
system of inequalities, 319

tan X, 236
trapezoid, 259
tree diagram, 141
trigonometric ratios, 236

undefined, 122
upper extreme, 18

variable, 2, 102
vector, 217
verify, 108
vertical line test, 171
volume, 258

x-intercept, 124

y-intercept, 117

hello my name is julia anne loxdale. I am bored in math class. its monday june 11 and on thursday is our final math exam. therefore this is Review Day. but I hate studying at school. I'm pretty sure it's impossible really. Derek and Johnny are being crazy as usual. oh dear. Becky hendry is being a substitute. she's fun fun fun. Right now I'm supposed to be studying. but... well this school year is almost over. it was interesting. I think next year will be enjoyable as well. ummmm... Tonite is my lifeguarding exam. wish me luck cuz I'm scared. to you dear student, I wish the greatest of luck in this math course. please let me know if you discover this note because I would like to laugh at my dumb self sometime in the future. well the ding dong is about to ding dong so farewell.

  ~Julia Loxdale.

     gr. 10   2006-2007

   intotheblucsea@gmail.com

  P.S. DON'T DO DRUGS

Matho matho matho!
I do not like the matho! I don't understand it,
I will never understand it. Booya!
It is very stressing. I donnot enjoy it
AT ALL!, I wonder how I will ever survive
3 years more of the matho!
Anyways I am Kim and yeah.
I have a fuzz......... and it is my head :)
My fav. colours is orange and lime green!
I'm going to be a Dental hygenist or a nanny OR a midwife.

I ♡ the Pit of ♡.

I ♡ balloons

I ♡ lunchtime.

I will one day be not famous and get married and have children and a garden, AND a car or something..
AND Puppies...Puppies that are eternally puppies!
AND I will live in the country and go barefoot and grow my fuzz long and go for walks at night.. but only when it is an enchanted moun.

⟩ ← enchanted moon (sliver moon)

OK back to the matho!

Oooh! AND I will drink tea! :) maybe.

2009-2010

P.S. I will also! Live on a sunflower farm and run through the fields of sunflowers every day!

340

Ahhh
weirdoweirdooo!